Environmental Emissions

Edited by Richard Viskup

Published in London, United Kingdom

IntechOpen

Supporting open minds since 2005

Environmental Emissions
http://dx.doi.org/10.5772/intechopen.90676
Edited by Richard Viskup

Contributors
David Galan Madruga, Rakhi Tailor, Yogesh Kumar Vijay, Minal Bafna, Maria Jedrusik, Dariusz Luszkiewicz, Arkadiusz Swierczok, Muhammad Zubair, Rabia Jathol Munsif, Ayesha Aziz, Muhammad Nadeem Zafar, Aniela Matuszewska, Maria Czaja, Farzaneh Hajirasouliha, Dominika Zabiegaj, Richard Viskup, Christoph Wolf, Werner Baumgartner, Olayemi Fehintola Awopeju

First published in London, United Kingdom, 2021 by IntechOpen
IntechOpen is the global imprint of INTECHOPEN LIMITED, registered in England and Wales, registration number: 11086078, 5 Princes Gate Court, London, SW7 2QJ, United Kingdom
Printed in Croatia

British Library Cataloguing-in-Publication Data
A catalogue record for this book is available from the British Library

Additional hard and PDF copies can be obtained from orders@intechopen.com

Environmental Emissions
Edited by Richard Viskup
p. cm.
Print ISBN 978-1-83968-510-1
Online ISBN 978-1-83968-511-8
eBook (PDF) ISBN 978-1-83968-512-5

For EU product safety concerns:
IN TECH d.o.o., Prolaz Marije Krucifikse Kozulić 3, 51000 Rijeka, Croatia, info@intechopen.com or visit our website at intechopen.com.

Contents

Preface

Today, the issue of environmental emissions is more important than ever before. Global emissions produced by humans and industries are increasing both linearly and exponentially due to the daily production, fabrication, and use of goods and commodities.

More and more goods are produced in semi-automated or fully automated production, and the types and quantities of fabricated items are expanding. However, these items have shorter lifespans than those fabricated 40 or 50 years ago. Nowadays it is common to replace appliances, electronics, and gadgets with newer models every few years. As such, factories are progressively producing more goods and thus producing more waste and emissions.

Although the global population is aging and the number of families and family members in households are decreasing, overall the population is still increasing, especially in cities. More people equals more stuff, and humans' demand for commodities is having negative results on the environment. Global natural resources and energy are being depleted. Growing numbers of people are driving or taking public transport daily and thus traffic is increasing, as is the number of vehicles. Over the last 30 years, transportation and logistics services have grown at rapid rates.

Today, almost every single item we bring into our homes is manufactured somewhere else and transported to us using trucks, ships, airplanes, trains, cars, and so on. Apart from foodstuffs, many of these items will end up in the trash or, in the best-case scenario, recycled. Although recycling is the better option, this too involves transporting materials, which results in the production of waste and emissions. Air pollution with particulates, soot, carbon, aerosols, heavy metals, and so on is causing adverse effects on human health as well as the environment.

This book, Environmental Emissions, presents new research and findings related to environmental emissions, pollution, and future sustainability. The first section deals with health effects, the second section discusses emission monitoring and mitigation, and the third section covers emission composition and measurement. Chapters are written by researchers, scientists, and engineers in environmental science, combustion, industrial pollution, health, emission monitoring, emission mitigation, and air quality measurements.

I would like to thank all of the authors for their rigorous preparation of the individual book chapters, and I hope the readers will find this volume a useful and interesting resource.

RnDr. Dr. Eng. Richard Viskup, MSc., MPhil., PhD.
Johannes Kepler University Linz,
Austria

Section 1

Health Effects of Environmental Emissions

Chapter 1

Effects of Environmental Emissions on the Respiratory System: Secrets and Consequences

Farzaneh Hajirasouliha and Dominika Zabiegaj

Abstract

Human health has been affected adversely by air pollution as a serious environmental challenge. Ambient (outdoor) air pollution mainly resulted from human activities (e.g., fuel combustion, heat generation, industrial facilities) causes 4.2 million deaths every year. Moreover, each year, 3.8 million people die from indoor air pollution which means household exposure to smoke from fuels and dirty cook stoves. They are the risks of stroke, heart attack, lung disease, or cancer that resulted from air pollution which assaults our brain, heart, and lungs using its invisible weapons named particulate matter (PM). These inhalable particles are of a nanoscale or microscale size. Upon inhalation, the air with its components enters the human body through the respiratory system. The lungs are the responsible organs for gas exchange with blood. Inhaled particles, such as silica, organic compounds, and metallic dusts, have toxic effects on our pulmonary system. For example, the accumulation of nanoparticles in the kidneys, liver, spleen, and central nervous system through the penetration of the epithelial barriers in the lungs has been observed. The purpose of this chapter is to describe the toxic effects of air particles on the different organs in the human body and to introduce some of the adverse effects of air pollution on human health.

Keywords: pulmonary system, human health, toxicity, nanoparticles, PM2.5

1. Introduction

Environmental emission is the pollutants around us in a gas or vapor phase. They can be gas or even solid particles in the air we breathe. If these materials find a way to get in touch with our human body, they may cause adverse effects on our health. There are various ways that they can reach human organs, from the skin to internal organs such as the lungs and brains. The toxicity of these particles depends on the nature of the particles and their size. In this chapter, the size of the particles that are the most dangerous ones is introduced, and the diseases and some of the mechanism of actions of these pollutants will be discussed.

2. Environmental emissions and human health

Environmental emissions are the pollutants discharged to our surrounding environment in a gas or vapor phase. Therefore, the outspoken result of such extensively spread emissions is air pollution.

According to some similarities among the pollutants, they can be classified in four groups:

1. Gaseous pollutants (e.g., nitrogen oxides, carbon monoxide, volatile organic compounds, ozone, sulfur dioxide)

2. Persistent organic pollutants (e.g., pesticides and dioxins)

3. Heavy metals (e.g., mercury, lead, chromium, vanadium)

4. Particulate matters (PMs) [1]

Air pollution causes 7 million deaths worldwide every year. Ambient air pollution mostly arising from anthropogenic activities [2, 3], e.g., using vehicles, combustion of fossil fuels, and power generation, is the cause of 4.2 million deaths due to acute and chronic effects on human health [4].

The diseases and, consequently, the deaths because of air pollutants are related to the routes of exposure to them. Skin diseases, lung cancer, and strokes are the examples of these illnesses.

The various adverse effects of air pollutants on the human health will be explained in the following sections.

3. Skin diseases

The skin is one of the first barriers against air pollutants. However, if this biological shield is exposed to air particulate matters for a long time, several types of diseases may happen to this largest organ in the human body. Atopic dermatitis, eczema, acne, and psoriasis are some of these diseases [5].

When toxic compounds are absorbed through the skin, they can cause local or systematic toxicity. In fact, the skin is an entrance for PMs to penetrate into the body through transfollicular route.

The particles with a diameter of 2.5 microns or less (PM2.5) and nitrogen dioxide can be effective on eczema and allergic sensitization. Particles with a diameter equal to or less than 10 μm (PM10) along with PM2.5 and ultrafine particles intensify itching as a symptom of atopic dermatitis. Skin aging through the process of extrinsic aging is another skin-related consequence of environmental pollution. In fact, environmental factors that cause the release of harmful free radicals will result in coarse wrinkles and uneven pigmentation of the skin. Skin cancer is the other disease caused by air pollutants such as polycyclic aromatic hydrocarbons as a potential group of carcinogenic materials. There are two potential routes for air particulate matters to diffuse through the skin surface: one of them is hair follicles or sweat ducts and the other one is across the stratum corneum. The skin barrier is degraded by PM2.5 because these particles reduce the levels of filaggrin, cytokeratin, E-cadherin, and tight junction molecules [6].

By interaction between aryl hydrocarbon receptor (AhR), as a receptor, and these air particulate materials, carcinogenic metabolites are generated, and the carcinogenicity in the cells will be induced [7]. It has been shown that there is no carcinogenic effect on the skin of the mice with AhR-deficiency. Moreover, in AhR-positive mice cases, the incidence of squamous cell carcinoma by air particulate matters was detected.

In fact, AhR is a ligand-activated transcription factor, and it has been reported that it is involved in maintaining cellular homeostasis. AhR has been recognized as

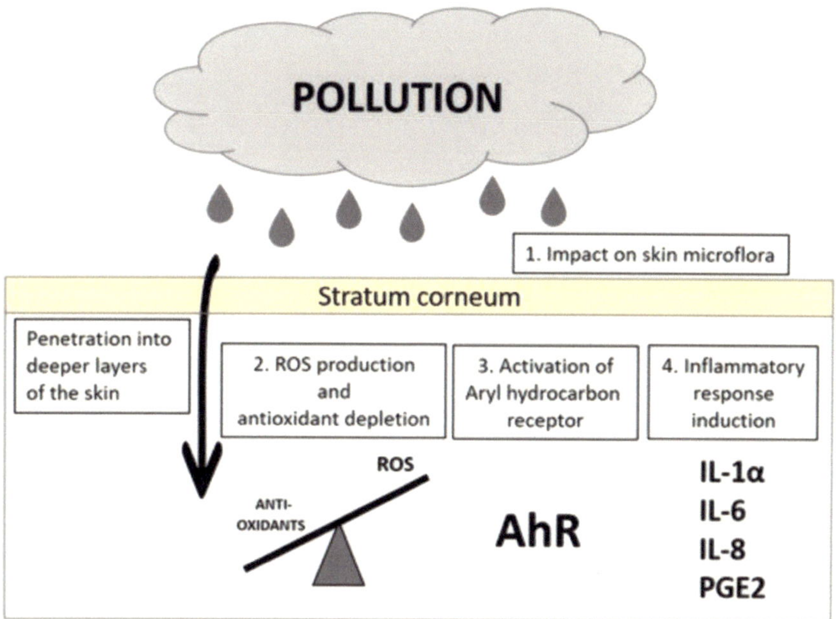

Figure 1.
Mechanism of the actions of pollutants on the skin [12].

a receptor for environmental pollutants, such as polycyclic aromatic hydrocarbons (PAHs). Upon ligand interactions, AhR in the cytoplasm will translocate into the nucleus and will bind to specific regulatory DNA sequences called dioxin response elements (DREs) which have been located in the promoters of target genes. The mentioned target genes are the ones involving detoxification the enzymes such as CYP1A1 and CYP1B1 [8]. Therefore, AhR is a regulator in controlling the CYP1 gene expression [9].

CYP1A1 is one of the cytochrome P450 enzymes that is capable to activate compounds with carcinogenic characteristics [10]. This enzyme can biotransform PAHs to carcinogens [11].

In fact, as **Figure 1** shows, pollution affects the skin microflora, and the pollutants will pass through the stratum corneum layer of the skin. Then, the reactive oxygen species (ROS) will be produced that causes a depletion in the amount of antioxidants in the skin. AhR will be activated, and the overproduction of pro-inflammatory factors will happen. These factors, as the indicators of cell response to the air pollutants, are the inflammation markers, such as prostaglandins (PGE2), interleukins (IL-6, IL-8, IL-1α, and IL-1β), or tumor necrosis factor-α (TNF-α). This mechanism will affect the biological function of the cells in the skin which will result in skin lesions and deterioration of the skin appearance [12, 13].

There are different studies on specific populations, e.g., asphalt-paving workers, chimney sweeps, coke oven workers, and asphalt-roofing workers, which show that skin uptake is a direct route of contamination by pollutants [14].

4. Heart diseases

The main morbidity and mortality caused by air pollutants is because of their adverse effects on cardiovascular system. **Figure 2** also demonstrates the deaths attributable to PM2.5 by the type of disease that it has caused between 1990 and 2015.

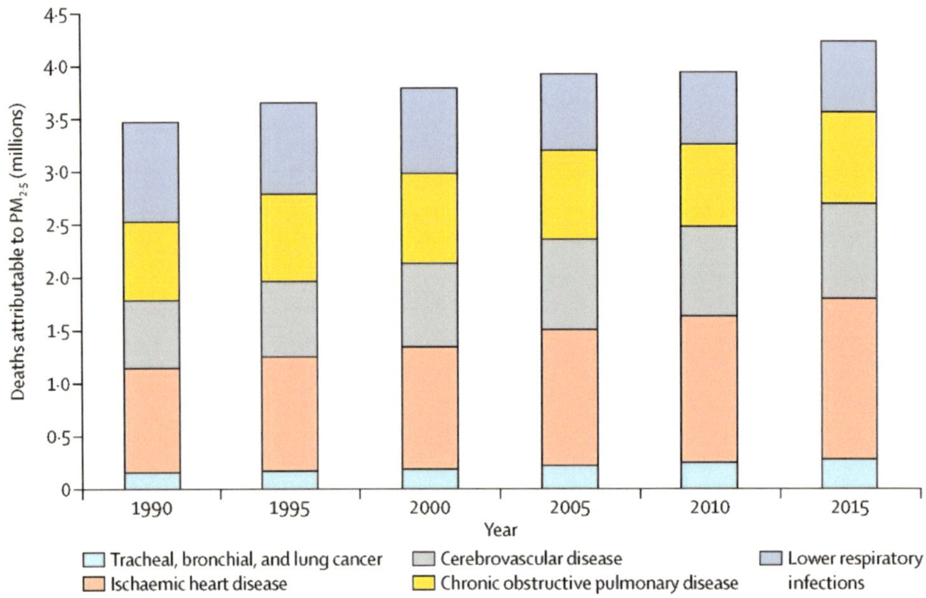

Figure 2.
Deaths attributed to ambient particulate matter pollution according to the type of diseases between 1990 and 2015 [16].

Epidemiological studies have shown that there is a robust association between air pollution and cardiovascular diseases. The Global Burden of Diseases (GBD) study estimated that the air pollution was the cause of 19% of all cardiovascular deaths in 2015 [15, 16].

A study on healthy people who spent 5 days near a steel plant showed that both of the systolic and diastolic blood pressures were higher in those volunteers who did not use mask in this area [17].

Increasing the exposure to the particulate matters will cause a promotion in the relative risk of cardiovascular diseases. A short-term increase in PM2.5 elevates the risk of acute cardiovascular events by 1–3% in a few days. This risk will be increased by almost 10% over several years which means long-term exposures. According to the World Health Organization air quality guidelines, the standard levels of exposure must be <20 μg/m^3 for daily levels and <10 μg/m^3 for annual levels. However, it has been shown that more than 90% of the global population is exposed to the levels exceeding these standard levels.

The secondary pathways through which the air pollution causes the risk of cardiovascular diseases can be classified into six groups:

1. Endothelial barrier disruption/dysfunction

2. Inflammation which involves both of the innate and adaptive immune components

3. Prothrombotic pathways

4. Autonomic imbalance

5. Central nervous system effects on metabolism and hypothalamic-pituitary-adrenal axis activation

6. Epigenomic changes

There are three primary pathways initiating those abovementioned secondary pathways. They are:

1. Oxidative stress and depletion of antioxidants

2. Direct translocation of particles and penetration into the systemic circulation

3. Biological intermediates through increasing oxidized by-products that may be involved in endothelial barrier dysfunction and inflammatory cell recruitment [18]

Exposure to PM2.5 can elevate the blood glucose level and incident cardiovascular disease events such as coronary artery disease [19–21].

The short-term exposure to particulate matters can cause acute cardiovascular diseases such as cardiac arrhythmia and myocardial infarction, while long-term exposure to these particles will result in coronary events [22].

5. Neurological diseases

The other adverse effect of air pollution is on the central nervous system (CNS). Air pollutants exacerbate the neurodegenerative conditions such as Parkinson's and Alzheimer's diseases. An association has also been shown between air pollution and the incidence of dementia, cognitive impairment, and white matter injury [23].

Increased concentration of traffic-related air pollution will increase Attention Deficit Hyperactivity Disorder (ADHD) and autism. It also affects adult cognition (episodic memory) and major depressive disorders [24].

The neuroinflammation and accumulation of B-amyloid peptide (AB42) and alpha-synuclein in the brain is the pathway that provides a potential mechanism for neurodegeneration [23].

There are two mechanisms through which air pollutants reach the brain. The first one is direct mechanism which means the pathways in which the adsorbed and soluble compounds reach the brain. The other mechanism is the peripheral one. Air pollution causes pro-inflammatory signals originating in the peripheral organs and/or tissues such as the lung, cardiovascular system, and liver. It increases the systemic-induced cytokine response which transfers the inflammation to the brain. TNF-α and IL-1β are the circulating cytokines which cause neuroinflammation, cerebral vascular damage, and neurotoxicity. Neuron damage is also caused by systemic inflammation.

Microglia, as the resident innate immune cells in our brain, are activated in neurodegenerative diseases such as Alzheimer's disease and Parkinson's disease. It has been reported that microglia activation happens by air particulate matters, e.g., manganese and titanium nanoparticles with reactive oxygen species.

Brain lipid peroxidation is a result of acute or chronic ozone exposure which can be the inhalation of reactive oxygen species [25–27].

6. Stroke

Stroke can be classified as both a neurological and cardiovascular disorder [28, 29]. As a cardiovascular disease, some mechanisms have been suggested. For example, air particles can be associated with an increase in the plasma lipoprotein-associated phospholipase A2 which has been proved to be an independent risk factor for stroke. Atrial fibrillation is also a known risk factor for stroke. Some studies on animals have shown that exposure to particles increases the incidence or susceptibility to arrhythmia. Therefore, it can also be considered as a mechanism to explain the observed link between air pollution and cardioembolic strokes [15]. Some molecular and cellular mechanisms of neuronal injury induced by air pollution have been suggested. Using these mechanisms, the association between air pollution and stroke as a neurological disorder can be explained. The blood-brain barrier (BBB), as the major site of controlled blood-CNS exchange, is a physical barrier which protects the CNS from potential pathogenic agents and toxins. So, an intact BBB is essential for the CNS to operate properly. It has been shown that the integrity of BBB is impaired in CNS diseases such as stroke, Alzheimer's disease, and Parkinson's disease. Activation or damage of the different cellular components of BBB results in CNS injury. When air particles are inhaled, systemic inflammation can be induced by them. This effect can disturb the integrity of BBB which increases the risk of stroke [30].

It has been shown that PM2.5 is associated with stroke mortality [31, 32]. The danger of air pollution in this case is as high as it is called a silent killer that needs environmental and public health policies [33, 34].

After exposure to air pollutant, the risk of ischemic stroke will be elevated [35]. Short-term exposure to pollution is considered to be associated with cardioembolic stroke. During peak periods of pollution, special care should be taken for those people susceptible to cerebral embolism [36]. So, cardiovascular diseases as a result of air pollution can also result in stroke. Carbon monoxide, nitrogen dioxide, and sulfur dioxide increase the risk of stroke [37, 38].

7. Pulmonary diseases

Upon inhalation, the air with its components enters the human body through the respiratory system. Based on the function, the respiratory system can be divided into two zones: conducting zone (nose to bronchioles) and respiratory zone (alveolar duct to alveoli). Anatomically, the respiratory tract is classified to the upper respiratory tract (organs outside the thorax—nose, pharynx, and larynx) and the lower respiratory tract (organs within the thorax—trachea, bronchi, bronchioles, alveolar duct, and alveoli) which makes up the lungs [39]. The extra-thoracic region, in the upper respiratory tract, includes the nasal and oral passages, pharynx, and larynx. This is the first line of defense against inhaled particles [40]. The tracheobronchial and alveolar regions are in the lower respiratory tract (**Figure 3**).

The lungs are the responsible organs for gas exchange with blood. This function is performed in alveolar sacs called alveoli located in the deepest region of the lung or alveolated region [43]. In fact, when we breathe, the air reaches the alveolar region (parenchyma) through a conducting airway tree. As the inset in **Figure 3** shows, blood flows in a capillary network in inter-alveolar septa. The last barrier that air and its particles encounter, before entering blood circulation, in the respiratory tract consists of a continuous alveolar epithelium, a continuous capillary endothelium, and a connective tissue layer between them [44].

Type II cells in the alveolar epithelium secrete a material which profoundly decreases the surface tension of the alveolar lining fluid. The important constituent of this surfactant is dipalmitoylphosphatidylcholine (DPPC) [45].

When we inhale the polluted air, a lot of particulate matter enters our respiratory system. However, the deposition of particles in various regions of our pulmonary system is dependent on the size of the particles. **Figure 4** demonstrates the fraction of particle deposition in the pulmonary system based on the particle size.

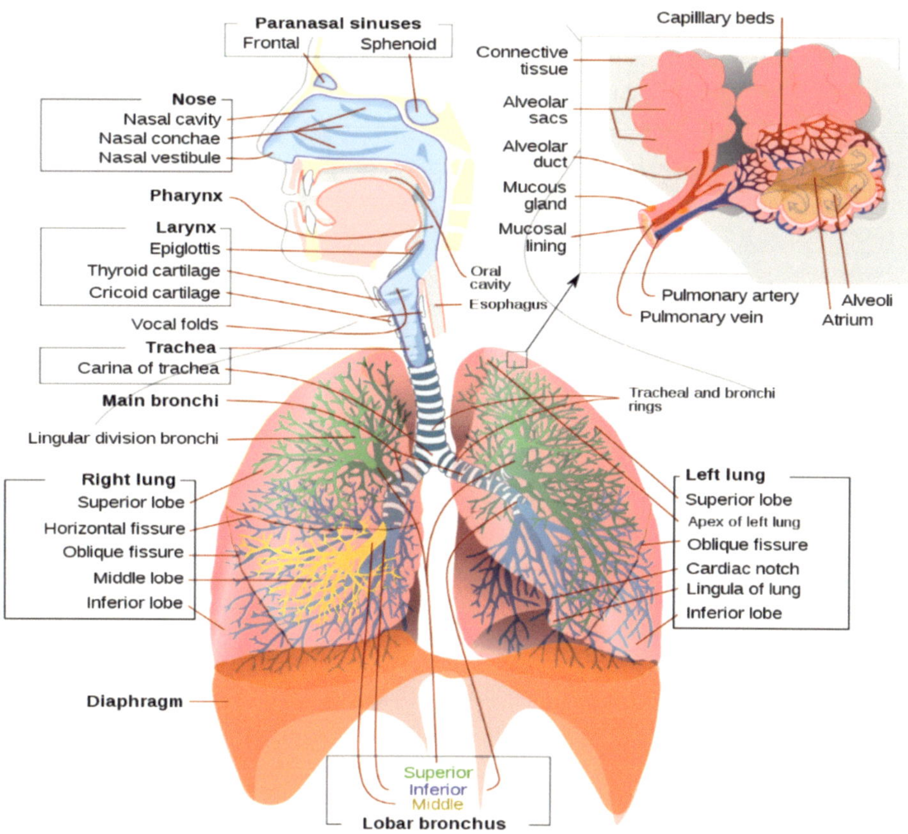

Figure 3.
Human respiratory system [41]. At the end of the branches of the airway tree, alveolar region exists. It contains the alveoli sacs. The gas exchanges between alveolar air and blood of the pulmonary capillaries [42].

In each region of the respiratory tract, various amounts of a certain size of particles (nanoparticles and microparticles) are deposited. For example, the highest deposition in the alveolar region is related to 20-nm and 3-μm particles. The smallest (1-nm) and largest (10-μm) particles are mostly deposited in the extra-thoracic region which includes the nasal and oral passages, pharynx, and larynx. This is the first line of defense against inhaled particles.

Air pollutants, after entering and passing the airways in the pulmonary system, will be changed into a danger for human health. Those particles able to diffuse through the lung barrier and enter the blood circulation will cause perturbation in normal functions in the human body. For example, blocking the vessels causes blood clotting, and consequently, the stroke will happen.

Inhaled particles, such as silica, organic compounds, and metallic dusts, have toxic effects on the pulmonary surfactant [47]. For example, the accumulation of nanoparticles in the kidneys, liver, spleen, and central nervous system through the penetration of the epithelial barrier of the alveolar region has been observed [48, 49].

As it was mentioned above, PM2.5 which also includes nanoparticles can penetrate the lung barrier. In fact, these particles are those which can reach the alveolar region, contact the lung surfactant, and interact with DPPC [50, 51]. Ultrafine particles of diesel exhaust and insoluble silicate particles of micrometer size can adsorb the components of lung surfactant and, therefore, be affected in terms of their in vitro expression of genotoxicity or cytotoxicity [52].

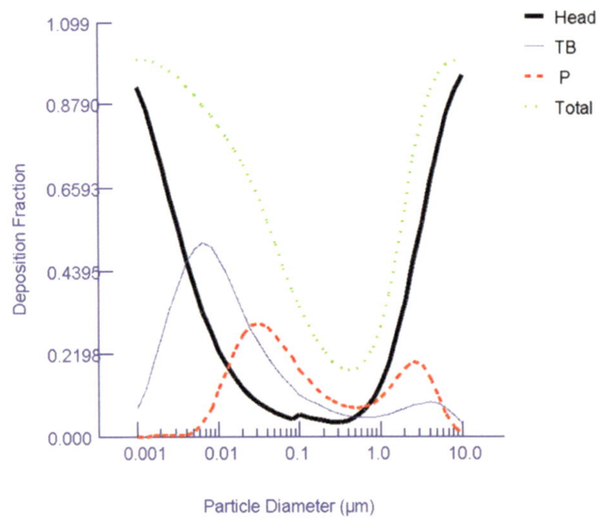

Figure 4.
Deposition fraction of particles in the various regions of the respiratory system according to the particle size. Head, TB, and P are the representatives of extra-thoracic, tracheobronchial, and alveolar regions, respectively. Total means the overall deposition in the three regions (https://www.ara.com/products/multiple-path-particle-dosimetry-model-mppd-v-304) [46].

Nanomaterials are the particles with at least one dimension less than or equal to 100 nm. Although there are a lot of beneficial nanotechnology-based products developed continuously, the number of efforts for the assessment of hazards of the nanoparticles when released in the environment is limited [53, 54].

The National Institute for Occupational Safety and Health (NIOSH) of the USA has identified the toxicity of the nanoparticles as one of the top 10 critical topic areas which should be considered in addressing knowledge gaps. The points below are two concepts classified in this topic:

1. Physicochemical properties (e.g., size, solubility, shape) of nanoparticles because these properties affect the potential toxicity and cytotoxicity of these particles

2. Short-term and long-term impact of nanoparticles on the human body (e.g., lung) [53]

There is an association between lung cancer incidence and PM2.5 air pollution [55, 56]. The International Agency for Research on Cancer (IARC) classified airborne particulate matter and outdoor air pollution as carcinogenic to humans [57]. Asthma and chronic obstructive pulmonary disease (COPD) are the other lung diseases caused by ambient air pollution [58, 59]. There are also some effects of air pollution on pregnancy outcomes, for example, decreased neurological development, chromosomal aberrations casing teratogenic effects, low birth weight, and respiratory syndrome likewise bronchiolitis/bronchitis [60].

8. Conclusions

The human health is threatened by air pollution. When air particulate matters get in contact with our body, they cause adverse effects such as lung diseases, cardiovascular diseases, stroke, and skin and neurological diseases. There are diverse mechanisms through which these particles cause different diseases. Some of these diseases are caused directly by the adsorption of particles onto the organs and tissues, e.g., lungs, in the human body, and some others are originating from the peripheral circulation which transfers cytokines to the organs such as the brain. Seven million deaths per year resulted from air pollution clearly shows that this is a very significant issue that must be tracked and investigated not only in terms of health and medical effects but also by making a decision on public health policies and environmental regulations. Therefore, it is a requirement for responsible authorities to consider these comprehensive harmful effects of particulate matters and, in general, environmental emissions on human health.

Acknowledgements

The authors would like to address special acknowledgement to Northumbria University, Newcastle upon Tyne, United Kingdom, and express gratitude to their colleagues within Smart Materials and Surfaces Laboratory in the Faculty of Engineering and Environment. Thank you for your contribution in sharing scientific documents, numerous discussions, and support during the preparation of this book chapter.

Conflict of interest

The authors declare no conflict of interest.

Notes

This book chapter is an introduction to the adverse effects of air pollution on human health. So, it can be used as a reference which addresses the reasons why air pollution should be considered as a serious challenge, and the related research must be conducted to investigate these effects and the solutions required.

Abbreviations

PM	particulate matter
PM2.5	particles with a diameter of 2.5 microns or less
PM10	particles with a diameter of 10 microns or less
AhR	Aryl hydrocarbon receptor
PAHs	polycyclic aromatic hydrocarbons
DREs	dioxin response elements
CYP1A1	cytochrome P450 1A
CYP1B1	cytochrome P450 1B1
PGE2	prostaglandins
IL-6	interleukin 6
IL-8	interleukin 8
IL-1α	interleukin 1 alpha
IL-1β	interleukin 1 beta
TNF-α	tumor necrosis factor-α
GBD	Global Burden of Diseases

CNS	central nervous system
ADHD	Attention Deficit Hyperactivity Disorder
AB42	B-amyloid peptide
HRV	heart rate variability
BBB	blood-brain barrier
DPPC	dipalmitoylphosphatidylcholine
NIOSH	National Institute for Occupational Safety and Health
IARC	International Agency for Research on Cancer
COPD	chronic obstructive pulmonary disease

Author details

Farzaneh Hajirasouliha* and Dominika Zabiegaj
Faculty of Engineering and Environment, Northumbria University,
Newcastle upon Tyne, UK

*Address all correspondence to: farzaneh.hajirasouliha@northumbria.ac.uk

References

[1] Kampa M, Castanas E. Human health effects of air pollution. Environmental Pollution. 2008;**151**(2):362-367

[2] Kukutschová J et al. On airborne nano/micro-sized wear particles released from low-metallic automotive brakes. Environmental Pollution. 2011;**159**(4):998-1006

[3] Malachova K et al. Toxicity and mutagenicity of low-metallic automotive brake pad materials. Ecotoxicology and Environmental Safety. 2016;**131**:37-44

[4] World Health Organization. Air Pollution. Available from: https://www.who.int/health-topics/air-pollution#tab=tab_1

[5] Drakaki E, Dessinioti C, Antoniou CV. Air pollution and the skin. Frontiers in Environmental Science. 2014;**2**:11

[6] Peng F et al. Potential role of PM2. 5 in melanogenesis. Environment International. 2019;**132**:105063

[7] Kim KE, Cho D, Park HJ. Air pollution and skin diseases: Adverse effects of airborne particulate matter on various skin diseases. Life Sciences. 2016;**152**:126-134

[8] Huang S-K et al. Mechanistic impact of outdoor air pollution on asthma and allergic diseases. Journal of Thoracic Disease. 2015;**7**(1):23

[9] Chang TK et al. Inhibition of procarcinogen-bioactivating human CYP1A1, CYP1A2 and CYP1B1 enzymes by melatonin. Journal of Pineal Research. 2010;**48**(1):55-64

[10] Androutsopoulos VP, Tsatsakis AM, Spandidos DA. Cytochrome P450 CYP1A1: Wider roles in cancer progression and prevention. BMC Cancer. 2009;**9**(1):187

[11] Walsh AA, Szklarz GD, Scott EE. Human cytochrome P450 1A1 structure and utility in understanding drug and xenobiotic metabolism. Journal of Biological Chemistry. 2013;**288**(18):12932-12943

[12] Rembiesa J et al. The impact of pollution on skin and proper efficacy testing for anti-pollution claims. Cosmetics. 2018;**5**(1):4

[13] Dobreva ZG et al. Proinflammatory and anti-inflammatory cytokines in adolescents from southeast Bulgarian cities with different levels of air pollution. Toxicology and Industrial Health. 2015;**31**(12):1210-1217

[14] Araviiskaia E et al. The impact of airborne pollution on skin. Journal of the European Academy of Dermatology and Venereology. 2019;**33**(8):1496-1505

[15] Lee KK, Miller MR, Shah AS. Air pollution and stroke. Journal of Stroke. 2018;**20**(1):2

[16] Cohen AJ et al. Estimates and 25-year trends of the global burden of disease attributable to ambient air pollution: An analysis of data from the global burden of diseases study 2015. The Lancet. 2017;**389**(10082):1907-1918

[17] Kumarathasan P et al. Cardiovascular and inflammatory mechanisms in healthy humans exposed to air pollution in the vicinity of a steel mill. Particle and Fibre Toxicology. 2018;**15**(1):34

[18] Rajagopalan S, Al-Kindi SG, Brook RD. Air pollution and cardiovascular disease: JACC state-of-the-art review. Journal of the American College of Cardiology. 2018;**72**(17):2054-2070

[19] Erqou S et al. Particulate matter air pollution and racial differences

in cardiovascular disease risk. Arteriosclerosis, Thrombosis, and Vascular Biology. 2018;**38**(4):935-942

[20] Dabass A et al. Systemic inflammatory markers associated with cardiovascular disease and acute and chronic exposure to fine particulate matter air pollution (PM2. 5) among US NHANES adults with metabolic syndrome. Environmental Research. 2018;**161**:485-491

[21] McGuinn LA et al. Fine particulate matter and cardiovascular disease: Comparison of assessment methods for long-term exposure. Environmental Research. 2017;**159**:16-23

[22] Fiordelisi A et al. The mechanisms of air pollution and particulate matter in cardiovascular diseases. Heart Failure Reviews. 2017;**22**(3):337-347

[23] Babadjouni RM et al. Clinical effects of air pollution on the central nervous system; a review. Journal of Clinical Neuroscience. 2017;**43**:16-24

[24] Sram RJ et al. The impact of air pollution to central nervous system in children and adults. Neuroendocrinology Letters. 2017;**38**(6):389-396

[25] Block ML, Calderón-Garcidueñas L. Air pollution: Mechanisms of neuroinflammation and CNS disease. Trends in Neurosciences. 2009;**32**(9):506-516

[26] Maher BA et al. Magnetite pollution nanoparticles in the human brain. Proceedings of the National Academy of Sciences. 2016;**113**(39):10797-10801

[27] Lucchini R et al. Neurological impacts from inhalation of pollutants and the nose–brain connection. Neurotoxicology. 2012;**33**(4):838-841

[28] Neurological Disorders. Public Health Challenges. 2006. Available from: https://www.who.int/mental_health/neurology/neurological_disorders_report_web.pdf

[29] Cardiovascular Diseases (CVDs). 2017. Available from: https://www.who.int/news-room/fact-sheets/detail/cardiovascular-diseases-(cvds)

[30] Genc S et al. The adverse effects of air pollution on the nervous system. Journal of Toxicology. 2012;**2012**:782462: 1-23

[31] Zhang R et al. Acute effects of particulate air pollution on ischemic stroke and hemorrhagic stroke mortality. Frontiers in Neurology. 2018;**9**:827:1-7

[32] Huang K et al. Long term exposure to ambient fine particulate matter and incidence of stroke: Prospective cohort study from the China-PAR project. BMJ. 2019;**367**:16720: 1-9

[33] Graber M et al. Air pollution and stroke. A new modifiable risk factor is in the air. Revue Neurologique. 2019;**175**(10):619-624

[34] Bejot Y et al. A review of epidemiological research on stroke and dementia and exposure to air pollution. International Journal of Stroke. 2018;**13**(7):687-695

[35] Ho AF et al. The relationship between ambient air pollution and acute ischemic stroke: A time-stratified case-crossover study in a city-state with seasonal exposure to the southeast Asian haze problem. Annals of Emergency Medicine. 2018;**72**(5):591-601

[36] Chung J-W et al. Air pollution is associated with ischemic stroke via cardiogenic embolism. Stroke. 2017;**48**(1):17-23

[37] Surendran A et al. Is air pollution a leading risk factor for stroke: A review.

International Journal of ChemTech Research. 2016;**9**(9):381-389

[38] Sun S et al. Short-term exposure to air pollution and incidence of stroke in the Women's Health Initiative. Environment International. 2019;**132**:105065: 1-7

[39] Patwa A, Shah A. Anatomy and physiology of respiratory system relevant to anaesthesia. Indian Journal of Anaesthesia. 2015;**59**(9):533-541

[40] Cheng YS. Aerosol deposition in the extrathoracic region. Aerosol Science & Technology. 2003;**37**(8):659-671

[41] Respiratory System. Available from: https://en.wikipedia.org/wiki/Respiratory_system

[42] Powers KA, Dhamoon AS. Physiology, Pulmonary, Ventilation and Perfusion. Treasure Island, FL: StatPearls Publishing; 2019

[43] Scarpelli EM. Physiology of the alveolar surface network. Comparative Biochemistry and Physiology Part A: Molecular & Integrative Physiology. 2003;**135**(1):39-104

[44] Knudsen L, Ochs M. The micromechanics of lung alveoli: Structure and function of surfactant and tissue components. Histochemistry and Cell Biology. 2018;**150**(6):661-676

[45] West JB. Respiratory Physiology: The Essentials. 8th ed. Philadelphia: Lippincott Williams & Wilkins, a Wolters Kluwer Business; 2008

[46] Miller FJ et al. Improvements and additions to the multiple path particle dosimetry model. Journal of Aerosol Science. 2016;**99**:14-26. Available from: https://www.ara.com/products/multiple-path-particle-dosimetry-model-mppd-v-304

[47] Griese M. Pulmonary surfactant in health and human lung diseases: State of the art. European Respiratory Journal. 1999;**13**(6):1455-1476

[48] Xu Y et al. Perturbation of the pulmonary surfactant monolayer by single-walled carbon nanotubes: A molecular dynamics study. Nanoscale. 2017;**9**(29):10193-10204

[49] Arick DQ et al. Effects of nanoparticles on the mechanical functioning of the lung. Advances in Colloid and Interface Science. 2015;**225**:218-228

[50] Guzmán E et al. Interaction of carbon black particles and dipalmitoylphosphatidylcholine at the water/air interface: Thermodynamics and rheology. The Journal of Physical Chemistry C. 2015;**119**(48):26937-26947

[51] Guzmán E et al. Influence of silica nanoparticles on phase behavior and structural properties of DPPC—Palmitic acid Langmuir monolayers. Colloids and Surfaces A: Physicochemical and Engineering Aspects. 2012;**413**:280-287

[52] Wallace WE, Keane MJ, Murray DK, Chisholm WP, Maynard AD, Ong T. Phospholipid lung surfactant and nanoparticle surface toxicity: Lessons from diesel soots and silicate dusts. In: Maynard AD, Pui DYH, editors. Nanotechnology and Occupational Health. Dordrecht: Springer; 2006. pp. 23-38

[53] NIOSH, T.N.I.f.O.S.a.H. 10 Critical Topic Areas. 2018. Available from: https://www.cdc.gov/niosh/topics/nanotech/critical.html [Accessed: 29 March 2018]

[54] Harishchandra RK, Saleem M, Galla H-J. Nanoparticle interaction with model lung surfactant monolayers. Journal of the Royal Society Interface. 2009;**7**(suppl_1):S15-S26

[55] Guo Y et al. The association between lung cancer incidence and ambient air

pollution in China: A spatiotemporal analysis. Environmental Research. 2016;**144**:60-65

[56] Čabanová K et al. Detection of micron and submicron particles in human bronchogenic carcinomas. Journal of Nanoscience and Nanotechnology. 2019;**19**(5):2460-2466

[57] Turner MC et al. Ambient air pollution and cancer mortality in the cancer prevention study II. Environmental Health Perspectives. 2017;**125**(8):087013:1-8

[58] Jacquemin B et al. Ambient air pollution and adult asthma incidence in six European cohorts (ESCAPE). Environmental Health Perspectives. 2015;**123**(6):613-621

[59] Long-Term Exposure to Air Pollution and Chronic Bronchitis. England: Public Health; 2016. p. 86. Available from: https://assets.publishing.service.gov.uk/government/uploads/system/uploads/attachment_data/file/541745/COMEAP_chronic_bronchitis_report_2016__rev_07-16_.pdf

[60] Kashyap R et al. A review on effect of air pollutants on fetal development and pregnancy outcomes. Journal of Air Pollution and Health. 2019;**4**(3):193-208

Chapter 2

Health Effect of Biomass Fuel Smoke

Olayemi Fehintola Awopeju

Abstract

Almost half of the world population rely on solid (biomass fuel and coal) for cooking, heating and lightning purpose. The resultant exposure to fine particulate matter from household air pollution is the seventh-largest risk factor for global burden of disease causing between 2.6 and 3.8 million premature deaths per year. The health effect ranges from cardiovascular, respiratory, neurocognitive and reproductive health effect. The most important are cardiovascular and respiratory health effects; others are the risk of burns and cataract in the eyes. Biomass fuel is any living or recently living plant and animal-based material that is burned by humans as fuels, for example, wood, dried animal dung, charcoal, grass and other agricultural residues. Biomass fuels are at the low end of the energy ladder in terms of combustion efficiency and cleanliness. Incomplete combustion of biomass contributes majorly to household air pollution and ambient air pollution. A large number of health-damaging air pollutants are produced during the incomplete combustion of biomass. These include respirable particulate matter, carbon monoxide, nitrogen oxides, formaldehyde, benzene, 1, 3 butadiene, polycyclic aromatic hydrocarbons (PAHs), and many other toxic organic compounds. In this article, health effects of biomass fuel use will be described in details highlighting the most affected systems and organs of the body.

Keywords: health effects, biomass fuel, household air pollution, ambient air pollution, particulate matter

1. Introduction

Almost half of the world populations rely on biomass fuels (BMF) for cooking, heating and lightning purpose. Household air pollution (HAP) from incomplete combustion of BMF is now understood to be a major risk factor for adverse health outcomes [1]. According to the 2016 Global Burden of Disease Study (GBD), HAP is ranked as the single most significant environmental health risk factor globally. It accounted for 2.6 million deaths and 77.2 million disability-adjusted life years (DALYs) in year 2016 [2], with greater than 99% of death occurring in low and middle income countries [3]. The health effect ranges from cardiovascular, respiratory, neurocognitive and reproductive health effect. The most important one is cardiovascular and respiratory health effect [4–6].

1.1 Historical considerations

The association between high levels of air pollutants and adverse health outcomes has been known since the seventeenth century [7]. In the 17th century, Queen Elizabeth I forbid the burning of coal near the palace at Westminster due to the unpleasant nature of smoke, by 1661 in London, John Evelyn suggested that factories should be located far from residential area [7]. In the 19th century, clinicians started linking lung diseases to air pollution in England [8]. Also, smog incidents in Meuse Valley, Belgium in 1930, Donora, Pennsylvania in 1948, and London, UK in 1952 acutely affected the elderly and those with existing cardiac and respiratory diseases and it caused increased hospitalizations and deaths. As a result of the London smog, an estimated 4000 people died and over 100,000 people suffered adverse health effects [9, 10]. Earlier this year, millions of Australians have been reported to be affected by air pollution, especially in the southwestern part of Australia where fire killed about 20 people including 3 volunteered fire fighters from the country wildfire [11].

2. What are biomass fuels?

Biomass fuel (BMF) is any living or recently living plant or animal-based material that is burned by humans as fuels, for example wood, dried animal dung, charcoal, grass and agricultural residue such as straw and sticks, dried leaves, twigs and wild grass [12, 13]. Although biomass fuel is primarily used by women in developing countries for domestic cooking and it is also used in developed countries primarily for the purpose of heating at homes, for example, 5% of household surveyed in Australia used woodstoves for indoor heating. BMF may also be chosen for cooking in developing countries because of the flavor they impact during cooking processes e.g. barbecues, smoked meat and wood-fired pizza [14]. There is also occupational exposure to BMF in developing countries, such occur as fire fighters [15]. In addition, air pollution from BMF also result from planned forest fires for agricultural practices during autumn and spring, and bushfires from countries with substantial parks and bush lands such as Canada and the USA during summer [14].

Although, incomplete combustion of BMF from cooking and heating result mainly in household air pollution (HAP), it is also an important contributor to ambient (outdoor) air pollution (AAP), accounting for an estimated 10–30% of ambient fine particulate matter (PM) [16], this is particularly so in developing countries. However, exposure to biomass PM is increasing in developed countries mainly from domestic heating purposes, increasing wild fires, and which can substantially contribute to ambient PM concentrations, particularly in winter months [13–16].

2.1 Components of BMF

The air pollutants from burning of BMF is numerous and has been shown to consist of 200 different compounds. Some of the pollutants are PM, carbon monoxide, sulphur and nitrogen oxides; organic compound like formaldehyde, acrolein, etc. [17–19]. The exact chemical composition of biomass smoke is dependent upon the fuel type, the temperature of burning, whether an open fire or free radicals incinerator is used, and local conditions (e.g., wind speed, humidity, indoor or outdoor fires) [14].

Although there are many pollutants, it is PM that have received most of the attention in scientific literature, on the basis of robust epidemiological, clinical and toxicological association between PM, and respiratory and cardiovascular diseases [19, 20].

PM components of air pollution are mixtures of solid, liquid and mixed phased particles suspended in air. It consists of carbonaceous particles with associated adsorbed organic chemicals and reactive metals. Common components of PM include nitrates, sulfates, PAH, endotoxin, and metals such as iron, copper, nickel, zinc, and vanadium [17]. PM is heterogeneous and variations in the characteristics of particles (e.g. particle size, surface area, and composition) (e.g. PAH, metal, and endotoxin content) released from different emission sources can influence the biological response [21]. The composition of PM to air pollution is highly dependent on season, density of sources and the specific technologies employed as well as meteorology and topography. In middle and low income countries, homes using BMF with poor designs that do not have flues or hood to take smoke out of the living area are often affected by the adverse health effects of HAP due to lack of ventilations [12].

PM can be classified based on aerodynamic diameter; this determine the site of deposition [22–24]. PM with a diameter of 0.1 μm or less are termed ultrafine PM and are deposited in the alveoli. While diameter of 2.5 μm or less are termed fine PM.

PM which are light and can remain suspended in air for longer periods and they are deposited throughout the respiratory tract, particularly in small airways and alveoli [24, 25]. These particles can be inhaled deep into the lungs, and have been linked to oxidative stress and inflammation induced damage of the respiratory system [26]. Coarse PM has an aerodynamic diameter of 2.5–10 μm and are deposited in large airways [27, 28]. The concentration of PM can be as high as 100 times the recommended 24 hour concentration by the U.S. Environmental Protection Agency and the WHO [29]. Although much of the research has been on PM, other components of BMF contribute significantly to the damaging effect of the respiratory system [30].

3. Health effect of particulate air pollution

Several studies have reported association of PM with different respiratory diseases, cardiovascular diseases, cancers, reproductive, neurocognitive and metabolic diseases [18, 25, 31, 32]. In a meta-analysis by Atkinson and colleagues, every 10 μg/cm3 increase in PM2.5 concentration was associated with a 1.04% (95% CI 0.52%–1.56%) increase in all-cause mortality [33]. Using 85 studies from 12 low- and middle-income countries in a meta-analysis study, Newell et al. reported that a 0.47% (95% CI 0.34–0.61) increase for cardiovascular mortality and 0.57% (95% CI 0.28–0.86) increase for respiratory mortality for every 10 μg/cm^3 increase in PM2.5 concentration [34].

3.1 Mechanisms

One of the commonly cited mechanism for the relationship between air pollutants is oxidative stress. Both particulate and gaseous pollutants can produce oxidative stress and can act independently or synergistically together [26]. However, most of the research on mechanism has focused on PM, as discussed earlier. Also, while the toxicological and health effects of air pollutants from BMF and other sources may be informative, people are exposed to a toxic mixture of all the components and maybe challenging to extrapolate individual effect to the compound exposure of air pollutants.

Oxidative stress refers to the imbalance between the productions of reactive oxygen species (ROS) and the cells ability to detoxify reactive intermediates or to repair cellular damage caused by ROS, examples of ROS are hydroxyl radicals and superoxide. When air pollution occurs, there is a dramatic increase in ROS level, resulting in significant damage to cellular components, including proteins, lipids, and DNA [35]. ROS can be derived from components of BMF smoke as well as the inflammatory cells recruited to the lungs [14]. Inflammatory response is produced through upregulation of pro-inflammatory cytokines such as tumour necrosis factor alpha, interleukin 6, and granulocyte colony stimulatory factor. Also, the immune cells are also recruited as well as upregulation of matrix metalloproteinases 2 and 9 and epithelial-mesenchymal transition [36]. In the airway and alveoli, this oxidative stress leads to alveolar macrophages activation and injury in the epithelial lining which in turn attracts inflammatory cells from the circulation [37]. In addition, PM in alveoli macrophages has been shown to modulate innate immune system and increased susceptibility to infection [38].

Lung inflammatory reaction can spill over into the systemic circulation and contribute to adverse effect in other organs [32]. PM exposure has been associated with systemic markers of oxidative stress which includes atherogenic precursors such as oxidized lipids, makers of hypercoagulability and thrombosis such as Von Willebrand factor and soluble CD 40 ligand, endothelial dysfunction, increased blood pressure and cardiac arrhythmias [31]. The systemic circulation could also be responsible for reproductive and intrauterine health effect [32].

Generation of ROS can also initiate free radical chain reactions which ultimately reach the nucleus and damage DNA leading to lung cancer and other cancers such as head and neck cancer. Additionally, gaseous pollutant present in air pollution such as nitrogen dioxide and volatile compound such as benzene can also lead to oxidative DNA damage. PAHs present in PM also form DNA adducts that has been implicated in carcinogenesis [39, 40].

4. Respiratory effects

There are compiling evidences associating exposure to solid fuel combustion products with respiratory diseases. Acute lower respiratory infection in children (ALRIs), chronic obstructive pulmonary disease (COPD) in women and lung cancer in women exposed to coal smoke are the three types of lung disease found to have strong evidence of association with exposure to solid fuel smoke: [12, 25].

4.1 Acute lower respiratory tract infection

The first report of indoor cooking smoke associated with childhood pneumonia and bronchiolitis was reported by Sofoluwe in Nigeria [41]. Acute lower respiratory tract infection (ALRI) is a leading contributor to the global burden of disease, it is also the commonest causes of morbidity and mortality particularly in children younger than five years. Almost all of this burden occurs in developing country where BMF is the primary source of household energy [12]. The relative risk for ALRIs for children exposed to BMF which include coal has been quantified in a number of studies [12]. In general, there is 2 to 3 times greater risk of developing ALRI in young children living in households exposed to solid fuel as compared to those not exposed [42].

The most recent meta-analysis by Smith and colleagues in 2014, using 23 observational studies and 1 randomized control trial, documented the pooled odds ratio of 1.78 (95% CI: 1.45–2.18) [18]. Although most of these studies are observational studies especially case control, and they used poor quality exposure measurement.

A recent study in Ethiopia, Adane and other investigators, recruited 5830 children less than 4 years old, and found that ALRI was linked with cow dung fuel use [AOR = 1.54 (95% CI: 1.02–2.330)], presence of extra indoors burning events

[AOR =2.19 (95% CI: 1.41–3.40)], child spending time near stove during cooking [AOR =1.41 (95% CI: 1.06–1.88)] and frequent cooking of meals [AOR =1.55 (95% CI: 1.13–2.13)] [43].

Although, there are many observational epidemiologic data that support an association between early childhood exposure to HAP and ALRI. Randomized control trials of interventions of improved cook stoves to reduce exposure to BMF smoke in order to prevent childhood ALRIs are emerging. A recent meta-analysis, of the six studies reporting child pneumonia outcomes, demonstrated no significant benefit in intention-to-treat analysis [44].

4.2 Tuberculosis

Tuberculosis is one of the leading causes of morbidity and mortality in the world, with a rate of 140 (95% CI: 915–1150) per 100,000 people. Exposure to HAP could impair the function of pulmonary alveolar macrophages and render the lungs prone to infections including tuberculosis [45]. There are conflicting results on the risk of tuberculosis and exposure to BMF. A meta-analysis by Lin et al. in 2014 reported that there were no association between BMF exposure and tuberculosis. With pooled odd ratio of 1.17 (95% CI: 0.83–1.65) for case–control study and 1.62 (95%CI 0.89–2.93) for cross sectional study [46]. Although, in Congo, Katoto reported that household air pollution is associated with chronic cough after completion of pulmonary tuberculosis treatment in adults [47].

4.3 COVID-19

The COVID-19 pandemic is still ongoing, there have been report of association of COVID-19 with AAP. Although, there are no available data specifically for HAP. Yao investigated the associations between PM concentrations and the case fatality rate of COVID-19 in 49 China cities, he found positive associations between PM pollution and COVID-19 in the 49 cities. Every increase in $PM_{2.5}$ and PM_{10} concentrations by 10 μg/m^3, raised the COVID-19 case fatality rate by 0.24% (0.01–0.48) and 0.26% (0.00–0.51) respectively [48]. Also, in Italy, Zoran and colleagues found positive correlations between confirmed COVID-19 daily new cases in Milan and air pollution with PM as follows, namely: with daily maxima PM_{10} (R^2 = 0.51), daily average surface air $PM_{2.5}$ (R^2 = 0.25), daily air quality index (R^2 = 4.35) [49]. In a USA study, after adjusting for factors such as population, weather, socioeconomic and behavioral variables, Wu and colleagues reported that increase of only 1 μg/m3 in $PM_{2.5}$ is associated with an 8% increase in the COVID-19 death rate (95% CI: 2%, 15%) [50]. However, air pollution has been documented to reduce during the lockdown period in different countries [51, 52], more epidemiological and experimental research are needed to estimate the impact of $PM_{2.5}$ on the incidence, severity and mortality of COVID-19.

4.4 Chronic obstructive pulmonary disease (COPD)

COPD is the fourth leading cause of mortality globally, causing more than 3 million death annually and over 80% of these deaths occur in low- and middle-income countries. It is also a substantial cause of economic and social burden [2]. It is characterised by persistent airflow limitation, associated with chronic inflammation of the airways and lungs in response to exposure to noxious particles and gases [53]. Previous systematic review has documented consistent association of biomass fuel with COPD. Kurmi et al. demonstrated that people exposed to BMF are at an increased risk of COPD compared to those that are not exposed [54]. Similarly, Hu et al. showed that people exposed to BMF smoke had increased odds ratio (OR) of

2.44 (95% CI, 1.9–3.33) for developing COPD, as compared to those that are not exposed, exposure to BMF was associated for developing COPD in Asian population, non-Asian population, in men and women [55].

In a recent meta-analysis by Sana et al., using 5 case–control studies and 19 cross sectional studies, they reported that biomass-exposed women were 1.38 times more likely to be diagnosed with COPD than non-exposed (OR 1.38, 95%Cl 1.28 to 1.57). The evidence is more for cross sectional studies than for case control studies but consistent in both rural and urban area [56]. HAP has been associated with increased risk of COPD exacerbation [57]. Effect of forest fire on emergency visit for COPD has been explored. Johnston et al. documented the odd ratio for COPD as 1.12 (95% CI: 1.02, 1.24) [58].

Intervention studies to prevent COPD from HAP is sparse, a prospective Chinese study show that substituting solid fuel with biogas for cooking and improving cooking ventilation were associated with a reduced decrease in FEV_1 and risk of COPD [59]. In a recent meta-analysis by Thakur et al., they documented that improved cook stoves were associated with a significant reduction in COPD among women: RR = 0.74 (95% CI 0.61 to 0.90). They also reported that ICS resulted in reductions in cough RR = 0.72 (95% CI 0.60 to 0.87), phlegm RR = 0.65 (95% CI 0.52 to 0.80), wheezing/breathing difficulty RR = 0.41 (95% CI 0.29 to 0.59) [60].

4.5 Asthma

Asthma is a non-communicable respiratory disease that is cause by chronic inflammation of the airways and results in wheezing, chest tightness, and cough. In 2015, approximately 400,000 people died of asthma worldwide [61]. In contrast to multiple studies on the risk of BMF smoke exposure and COPD, data are sparse on the risk of BMF and asthma. Although, there have been conflicting association or relationship between biomass exposure and asthma; evidences are emerging that biomass exposure may be linked with asthma risk, prevalence or incidence.

Barry et al. accessed 508 respondents and showed that individuals using wood or coal for cooking had increased odd of 2.3 (95% CI: 1.1–5.0) for reporting current asthma symptom [62]. Oluwole et al. also reported increased odds of asthma symptom in children who lived in household that used biomass had an adjusted OR of 1.33 (95% CI 1.05–1.6) for any of the asthma symptoms [63]. They further observed that biomass fuel use was associated with increased odds of severe symptoms of OR 2.39 (95% CI 1.16–4.84) [64]. Ayuk in their retrospective analysis of ISAAC phase III study found that open fire cooking was among the factors associated with asthma 1.28 (95% 1.06–1.51) [65].

However there are other studies that did not find an association between BMF and Asthma [66, 67]. Noonan et al. conducted a randomized control trial of air-filter intervention in asthmatic children exposed to BMF smoke, the intervention did not improve asthma quality of life. Although, there was an improvement in secondary measure of diurnal peak flow variability [68].

4.6 Other respiratory diseases

BMF exposure is also associated with interstitial lung disease referred to as 'hut lung' [12]. Hut lung disease is characterized by carbon disposition, dust macules, and mixed dust fibrosis, and it has been reported in cases primarily of women with chronic high-level exposure to indoor biomass smoke in developing countries [27]. In addition, bronchial anthracofibrosis has also been reported in elderly women who have worked long hours in poorly ventilated kitchen full of smoke due to incomplete combustion of BMF [69].

5. Cancers

5.1 Lung cancer

Lung cancer causes more death globally than any other cancer and it is the seventh leading cause of death globally [70], the International Agency for Research on Cancer concluded emissions from household coal combustion are a Group 1 carcinogen, while those from biomass were categorised as 2A due to epidemiologic limitations Although, smoking is the major risk factor for lung cancer worldwide, about 1.5% of lung cancer death are attributed to exposure to carcinogens from biomass fuel smoke annually [71].

In a meta-analysis to estimate the risk of lung cancer with the risk of BMF for cooking and heating, using 14 studies that were all case control, they found out that the risk of lung cancer with biomass for cooking and/or heating was OR 1.17 (95% CI 1.01 to 1.37). Although, more than 50% of the study did not report a clean reference category. When analyses restricted to studies with clean reference category, the evidence still remain the same for men and women [72]. A study published by Raspanti and colleagues, after the meta-analysis also reported OR: 1.77, (95% CI: 1.00–3.14), with the estimate more robust for non-smokers (P_{trend} = 0.01). Their study was a case control study of 606 lung cancer cases and 606 healthy controls matched on age (±5 years), gender, and geographical residence and adjusting for potential confounders such as tobacco use, tuberculosis status, Social economic status, age, gender, ethnicity, and exposure to second hand smoke. Conclusively, there was an increased risk of lung cancer among those who were exposed to HAPs [73].

5.2 Gastric cancer and esophageal cancer

Globally, gastric cancer is the fifth most common cancer and it is third among the causes of cancer mortality [74]. In a recent research carried out in Zambia, Kayamba et al. reported that there was an association between gastric cancer and reliance on BMF (OR, 3.5; 95% CI, 1.9 to 6.2; P < 0001) [40]. Okello and colleagues in a systematic review and meta-analysis using 16 case control studies, reported that the use of BMF was associated with increased risk of esophageal squamous cell carcinoma OR 3.02(95% CI 2.22, 4.11). Analysis by continent showed that Africa and Asia had the highest odds of esophageal squamous cell carcinoma [75].

5.3 Other cancers

Josyula and colleagues evaluated the relationship between HAP and other cancers apart from lung cancer in a meta-analysis, they found out that HAP was associated with cervical cancer neoplasia (OR = 6.46; 95% CI = 3.12–13.36); nasopharyngeal (OR = 1.80; 95% CI = 1.42–2.29); oral (OR = 2.44; 95% CI =1.87–3.19); laryngeal (OR = 2.35; 95% CI = 1.72–3.51) cancers and pharyngeal (OR = 3.56; 95% CI = 2.22–5.70). The association between HAP and upper aero-digestive cancers remained significant even when the analysis was restrained to only studies that controlled for smoking [76].

6. Cataract

Cataract is the clouding of the eye lens by preventing the passage of light and it is highly prevalent in developing countries. It is a leading cause of blindness globally [77]. Several studies have shown an association between cataract and BMF. GBD Risk factors collaboration in 2015 reported that cataract is a global leading cause of blindness and it account for approximately 0.12% of all DALYS [77]. Smith et al. using seven studies from India and Nepal provided estimates for association between HAP and cataract, they reported a pooled OR of 2.64 (1.74, 3.50); the evidence for men was inconclusive. However, the estimate for women was OR 2.47 (1.61, 3.73) was deemed reliable [14].

A meta-analysis also reported in 2014, despite study heterogeneity, BMF use was associated with an increased risk of cataract with summary effect size of 2.12; 95% CI 1.61–2.80 [78]. Thakur et al. in a meta-analysis of 3 studies documented that improved cook stoves among women significantly reduce the presence of ocular symptoms RR =0.58, (95% CI 0.43–0.78) [60].

7. Cardiovascular, cerebrovascular and metabolic diseases

Cardiovascular disease is a leading cause of mortality worldwide and this is rapidly increasing in developing countries [31]. Although, there is a growing body of research linking HAP with sub-clinical indicators of cardiovascular disease risk including blood pressure, carotid atherosclerotic plaque, and arterial stiffness, epidemiological evidence linking BMF smoke and cardiovascular disease is limited [32]. According to a recent publication, it was reported that in middle- and low-income countries, household air pollution, along with other factors had stronger effects on cardiovascular disease or mortality compared to high income countries [79]. According to WHO, 12% of all death due to stroke can be attributable to the daily exposure to household air pollution arising from cooking with solid fuels and kerosene [80].

Fatmi and Coggon conducted a meta-analysis of 26 studies, 10 in south Asia, 4 in China, 2 in Turkey, 1 in Iran and 8 in Central and south American reported that the current balance of epidemiological evidence points to an increased risk of cardiovascular disease from HAP as a consequences of using solid and especially BMF for cooking and heating. Relative risks from long term exposure could be 2- to 4- fold [81]. In accordance with this study, using nationally representative and internationally comparable data, Arku et al. examined the association between solid fuel use and BP in 77,605 largely premenopausal women (aged 15–49) from ten resource-poor countries. They found that primary use of solid fuel was associated with 0.58 mmHg higher systolic BP (95% CI: 0.23, 0.93) as compared to primary use of clean fuel [82]. Ofoli et al. reported that use of BMF was associated with higher systolic blood pressure, more carotid intima media thickness (CIMT) and increased odds of pre-hypertension (OR: 1.67, 95% CI: 1.57–4.99, P = 0.035) [83].

In a cross sectional Chinese study involving more than 14,000 men and women aged 18 years or older showed that BMF and coal was associated with self-report of physician diagnosed ischemic heart disease with an adjusted OR 2.6 for ever versus never of solid fuel and significant trend across duration of use for stroke, hypertension and diabetes [32].

In addition, Right heart failure and pulmonary arterial hypertension were associated with HAP. Interestingly, women were almost two times more likely to present with pulmonary arterial hypertension (OR 1.72, 95% CI 1.17–2.55; p = 0.006), suggesting gender- related risk factors that may include HAP and HAP was also identified as a risk factor in women with isolated right heart failure in Kenya [84].

Most of the studies on metabolic diseases arose from ambient air studies [31, 32]. For example, a large study conducted in the United States reported that diabetes prevalence increases by 1% with each 10 μg/m3 $PM_{2.5}$ [85]. Since HAP has been understudied with respect to cardiovascular disease most of our pathogenic hypothesis and more specific research is needed.

8. Reproductive and pregnancy health outcomes

There is limited but accruing evidence linking HAP exposure from solid fuel combustion with adverse maternal and perinatal outcomes. Also, epidemiologic evidence connecting AAP exposure with adverse pregnancy outcomes has been accumulating worldwide over the last two decades with several studies also attempting to summarize the available evidence [86].

In a meta-analysis by Amegah et al. including 19 studies, BMF use resulted in reduced birth weight, [86.4 g (95% CI: 55.5, 117.4)]. There was also an increased risk of low birth weight and stillbirth in those exposed to BMF, [OR (1.35, 95% CI: 1.23, 1.48)] and [OR (29, 95% CI: 1.18, 1.41)] respectively [86]. Interventional studies assessing HAP and low-birth weight are limited. However, Thompson et al. found an increase of 89 g in the birth weight of children of mothers using the intervention stoves (vs. open fires) (95% CI −27, 204), and reduced odds of a low birth weight child (OR 0.74, 95% CI 0.33–1.66) [87].

In a cross-sectional comparative study, Murkerjee and colleagues reported that there was positive association between BMF smoke exposure and menstrual irregularities such as irregular cycle (OR = 1.8, 95% CI 1.33–2.34), shortened menstrual cycle (OR = 5.1, 95% CI 3.62–9.21), spontaneous abortions (OR = 1.7, 95% CI 1.10–4.10), after controlling the potential confounders [88].

Agarwal and Yamamoto, using data from India's third National Family Health Survey (NFHS-3, 2005–2006) of 39,657 women aged 15–49 years who had a live birth in the previous 5 years, reported that women residing in houses using BMF had higher odd of reporting preeclampsia/eclampsia symptoms compared to those who reside in houses using cleaner fuels (OR = 2.21; 95%: 1.26–3.87; p = 0.006) [89]. This is further supported by a recent meta-analysis that showed maternal exposure to PM2.5 (per 10 $\mu g/m^3$ increment) is associated with elevated risk of preeclampsia (OR = 1.32, 95% CI 1.10 to 1.58%) especially in third trimester [90].

A recently published data from Ghana showed that using BMF was associated with adverse Apgar score at 5 min (aOR: 3.83, 95%CI: 1.44–10.11) and perinatal mortality (aOR: 7.6, 95%CI: 1.67–36.0) [91]. In sum, there are increasing evidences that BMF use are associated with menstrual irregularities, adverse pregnancy outcomes and perinatal morbidity and mortality.

9. Neurological health outcomes

There is a convergence of human, animal, and in vitro studies on the effects of air pollution on the brain, although most study has been associated with ambient PM [92]. Several neurological diseases have linked with exposure to air pollution ranging from neuro degenerative disease, to psychiatric disease, to neurodevelopmental and behavioural disorders in children.

Dementia denote memory loss and other cognitive abilities severe enough to interfere with daily life. Worse performance in tests evaluating episodic memory especially cognitive function tests had been associated with PM2.5 [93]. Other study has found associations between long-term exposure to PM2.5 among adults (>65 years) and neurodegenerative diseases such as Alzheimer's, and Parkinson's diseases [94].

Pre-natal maternal or child exposure to air pollutants during pregnancy, infancy or childhood (when the brain neocortex develops rapidly) has been related to delays in cognitive development in children [95, 96]. A metanalysis by Lam et al., involving 17 case–control, 4 ecological, 2 cohort studies, they documented that PM was associated with autism; summary odds ratios (ORs) of 1.07 (95% CI: 1.06, 1.08) per 10-μg/m^3 increase in PM_{10} exposure and 2.32 (95% CI: 2.15, 2.51) per 10-μg/m^3 increase in $PM_{2.5}$ exposure [97]. It has been postulated that air pollution induced oxidative stress can be related to dopaminergic neurotoxicity, and can cause depression and other neuropsychiatry disorder such as anxiety [93].

10. Conclusion

Exposure to air pollutants is one of the most important avoidable risks to health globally [98]. Air pollution has been termed the "silent killer" by the World Health Organization because its effects often go unnoticed or are not easily measured. Although the pulmonary and cardiac diseases have been the most studied adverse health, there are emerging evidence that air pollutants can damage any organ of the body and cause more ill-health. Chronic liver diseases [99], skin diseases [100], bone diseases [101] and autoimmune diseases [102] are increasingly been associated with air pollution. As data continues to increase on health effect of both HAP and AAP which are overlapping and contribute to each other, it is pertinent that all stakeholders; individuals, families, health care professionals, policy makers, governmental and non-governmental institutions work together to minimise the adverse health effects of the world's biggest environmental risk factor.

Acknowledgements

We appreciate Almighty God for granting the strength and wisdom to write the manuscript and we are also grateful to Miss Nene Raphael for typesetting this article.

Abbreviations

AAP	Ambient air pollution
ALRI	Acute lower respiratory tract infection
AOR	Adjusted odd ratio
BMF	Biomass fuel smoke
HAP	Household air pollution
CI	Confidence interval
COPD	Chronic obstructive pulmonary disease
COVID-19	Corona virus disease 2019
DALYs	Disability-adjusted life years
DNA	Deoxyribonucleic acid
FEV1	Forced expiratory volume in 1 second
GBD	Global Burden of Disease Study
OR	Odd ratio
P	Significance value
PAHs	Polycyclic aromatic hydrocarbons
$PM_{2.5}$	Particulate matter with aerodynamic diameter 2.5 μm or less
PM_{10}	Particulate matter with aerodynamic diameter 2.5–10 μm

PM	Particulate matter
R^2	Coefficient of determination
ROS	Reactive oxygen species
WHO	World health organisation
CD	Cluster of differentiation

Author details

Olayemi Fehintola Awopeju
Respiratory Unit, Department of Medicine, Obafemi Awolowo University Teaching Hospital, Ile Ife, Osun State, Nigeria

*Address all correspondence to: yemijide@yahoo.com

References

[1] World health organisation. Household air pollution and health. WHO, Geneva 2018. http://www.who.int/news-room/fact-sheets/detail/household-air-pollution-and-health.

[2] GBD 2016 Risk Factors Collaborators. Global, regional, and national comparative risk assessment of 84 behavioural, environmental and occupational, and metabolic risks or clusters of risks, 1990-2016: a systematic analysis for the Global Burden of Disease Study 2016. Lancet. 2017; 390(10100):1345-1422. DOI: 10.1016/S0140-6736(17)32366-8.

[3] Carlsten C, Salvi S, Wong GWK, Chung KF. Personal strategies to minimise effects of air pollution on respiratory health: advice for providers, patients and the public. Eur Respir J. 2020 4;55(6):1902056.

[4] Wu W, Jin Y, Carlsten C. Inflammatory health effects of indoor and outdoor particulate matter. The Journal of Allergy and Clinical Immunology. 2019;**141**:833-844

[5] Hoek G, Krishnan RM, Beelen R, et al. Long-term air pollution exposure and cardio-respiratory mortality: a review. Environmental Health. 2013;**12**:43

[6] Johnston HJ, Mueller W, Vardoulakis, et al. How harmful is particulate matter emitted from biomass burning? A Thailand perspective. Curr Pollution Rep. 2019;**5**:353-377

[7] Evelyn J. Fumifugium or the inconvenience of the air and smoke of London dissipated. Brighton, UK: National Society for Clean Air; 1661

[8] Brimblecombe P. London air pollution, 1500-1900. Atmospheric Environment. 1977;**11**:1157-1162

[9] Nemery B, Hoet PH, Nemmar A. The Meuse Valley for of 1930: an air pollution disaster. Lancet. 2001;**357**:704-708

[10] Logan WP. Mortality in the London fog incident, 1952. Lancet. 1953;**1**:336-338

[11] https://airqualitynews.com/2020/01/06/the-horror-of-the-australian-bush-fires-and-air-pollution

[12] Fullerton DG, Bruce N, Gordon SB. Indoor risk pollution from biomass fuel smoke is a major health concern in the developing world. Transaction of the royal society of tropical medicine and hygiene. 2008;**102**:843-851

[13] Bruce N, Perez-Padilla R, Albalak R. Indoor air pollution in developing countries :a major environmental and public health challenge. Bulletin of the World Health Organization. 2000;**78**:1078-1092

[14] Capistrano SJ, Reyk DV, Chen H, et al. Evidence of biomass smoke exposure as a causative factor for the development of COPD. Toxics. 2017;**5**:36

[15] Reisen F, Brown SK. Australian firefighters' exposure to air toxics during bushfire burns of autumn 2005 and 2006. Environment International. 2009;**35**:342-352

[16] Hoek G, Kos G, Harrison R, et al. Indoor–outdoor relationships of particle number and mass in four European cities. Atmospheric Environment. 2008;**42**:156-169

[17] Naeher LP, Brauer M, Lipsett M, et al. Wood smoke health effects: A review. Inhalation Toxicology. 2007;**19**:67-106

[18] Smith KR, Bruce N, Balakrishnan K, et al. Millions dead: how do we know and what does it mean? Methods used

in the comparative risk assessment of household air pollution. Annual Review of Public Health. 2014;**35**:185-206

[19] Sigsgaard T, Forsberg B, Annesi-Maesano I, et al. Health impacts of anthropogenic biomass burning in the developed world. Eur Respi J. 2015;**46**:1577-1588

[20] Brook RD. cardiovascular effects of air pollution. Clinical Science. 2008;**115**:175-187

[21] Kelly FJ, Fussell JC. Size, source and chemical composition as determinants of toxicity attributable to ambient particulate matter. Atmospheric Environment. 2012;**60**:504-526

[22] He C, Miljevic B, Crilley LR, et al. Characterisation of the impact of open biomass burning on urban air quality in Brisbane, Australia. Environment International. 2016;**91**:230-242

[23] Zelikoff JT, Chen LC, Cohen MD, Schlesinger RB. The toxicology of inhaled woodsmoke. J Toxicol Environ Health Part B Crit Rev. 2002;**5**:269-282

[24] Valavanidis A, Fiotakis K, Vlachogianni T. Airborne Particulate Matter and Human Health: Toxicological Assessment and Importance of Size and Composition of Particles for Oxidative Damage and Carcinogenic Mechanisms. Journal of Environmental Science and Health. 2008;**26**:339-362. DOI: 10.1080/10590500802494538

[25] Gordon SB, Bruce NG, Grigg J, et al. Respiratory risks from household air pollution in low and middle income countries. The Lancet Respiratory Medicine. 2014;**2**:823-860

[26] Laumbach RJ, Kipen HM. Respiratory health effects of air pollution: update on biomass smoke and traffic pollution. The Journal of Allergy and Clinical Immunology. 2012;**129**:3-11

[27] Sood A. indoor fuel exposure and lung in both developing and developed countries: An update. Clinics in Chest Medicine. 2012;**33**(4):649-665

[28] Guarnieri M, Balmes JR. Outdoor air pollution and asthma. Lancet. 2014;**383**:1581-1592

[29] WHO, International Programme on Chemical Safety/Air pollution. http://www.who.int/ipcs/features/chemicals_concern/en/.

[30] Assad NA, Vidit K, Sood A. Biomass smoke and chronic lung disease. Current Opinion in Pulmonary Medicine. 2016;**22**:150-157

[31] Hamanaka RB, Mutlu GM. Particulate Matter Air Pollution: Effects on the Cardiovascular System. Front. Endocrinol. 2018;**9**:680

[32] Balmes JR. household air pollution from domestic combustion of solid fuels and health. The Journal of Allergy and Clinical Immunology. 2019;**143**:1979-1987

[33] Atkinson RW, Kang S, Anderson HR, et al. Epidemiological time series studies of PM2.5 and daily mortality and hospital admissions: a systematic review and meta-analysis. Thorax. 2014;**69**:660-665

[34] Newell K, Kartsonaki C, Lam KBH, et al. Cardiorespiratory health effects of particulate ambient air pollution exposure in low-income and middle-income countries: a systematic review and meta-analysis. Lancet Planet Health 2017;1e:363-380.

[35] Genc S, Zadeoglulari Z, Fuss SH, et al. The adverse effects of air pollution on the nervous system. Journal of Toxicology. 2012;**782462**. DOI: 10.1155/2012/782461

[36] Rajendra KC, Shukla SD, Guatam SS, et al. Clin Trans

Med. 2018;7:39. DOI: 10.1186/s40169-018-0217-2

[37] Mondal NK, Saha H, Mukherjee B, et al. Inflammation, oxidative stress, and higher expression levels of Nrf2 and NQO1 proteins in the airways of women chronically exposed to biomass fuel smoke. Molecular and Cellular Biochemistry. 2018;**447**:63-76

[38] Lee A, Kinney P, Chillrud S, et al. a systemic review of innate immunomodulatory effects of household air pollution secondary to the burning of biomass fuels. Annals of Global Health. 2015;**81**:368-374

[39] Risom L, Moller P, Loft S. Oxidative stress-induced DNA damage by particulate air pollution. Mutation Research. 2005;**592**:119-134

[40] Kayamba V, Zyambo K, Mwakamui S, et al. Biomass smoke exposure is associated with gastric cancer and probably mediated via oxidative stress and DNA damage: a case control study. JCO Global Oncol. 2020;**6**:532-541

[41] Sofoluwe GO. Smoke pollution in dwellings of infants with bronchopneumonia. Archives of Environmental Health. 1968;**16**:670-672

[42] Smith KR, Samet JM, Romieu I, et al. Indoor air pollution in developing countries and acute lower respiratory infections in children. Thorax. 2000;**55**:518-532

[43] Adane MM, Alene GD, Mereta ST, et al. Prevalence and risk factors of acute lower respiratory infection among children living in biomass fuel using households: a community-based cross-sectional study in Nortwest Ethiopia. BMC Public Health. 2020;**20**(1):363. DOI: 10.1186/s12889-020-08515-w

[44] Saleh S, Shepherd W, Jewell C, et al. Air pollution interventions and respiratory health: a systematic review. The International Journal of Tuberculosis and Lung Disease. 2020;**24**:150-164

[45] Woldesemayat EM, Datiko DG, Lindtjorn B. Use of biomass fuel in household is not a risk factor for pulmonary tuberculosis in south Ethiopia. The International Journal of Tuberculosis and Lung Disease. 2014;**18**:67-72

[46] Lin HH, Suk CW, Lo HL, et al. Indoor air pollution from solid fuel and tuberculosis: a systemic review and meta-analysis. The International Journal of Tuberculosis and Lung Disease. 2014;**18**:613-622

[47] Katoto P, Murhula A, Kayembe-Kitenge T, et al. Household air pollution is associated with chronic cough but not haemoptysis after completion of pulmonary tuberculosis treatment in adults, rural eastern democratic republic of Congo. International Journal of Environmental Research and Public Health. 2018. DOI: 10.3390/ijerph15112563

[48] Yao Y, Pan J, Wang W, et al. Association of particulate matter pollution and case fertility of COVID-19 in 49 Chinese cities. Science of the Total Environment. 2020. DOI: 10.1016/j.scitotenv.2020.140396

[49] Zoran MA, Savastru RS, Savastru DM, et al. Assessing the relationship between surface levels of PM2.5 and PM10 particulate matter impact on COVID-19 in Milan, Italy. Science of the total environment 2020; DOI: 10.1016/j.scitotenv.2020.139825.

[50] Wu X, Nethery RC, Sabath BM, et al. Exposure to air pollution and COVID-19 mortality in the United States: A nationwide cross-sectional study. medRxiv 2020; DOI: 10.1101/2020.04.05.20054502.

[51] Kumar P, Hama S, Omidvarborna H, et al. Temporary reduction in fine particulate matter due to 'anthropogenic emissions switch-off' during COVID-19 lockdown in Indian cities. Sustainable Cities and Society. 2020. DOI: 10.1016/j.scs.2020.102382

[52] Zheng H, Kong S, Chen N, et al. Significant changes in the chemical compositions and sources of $PM_{2.5}$ in Wuhan since the city lockdown as COVID-19. Science of the total environment 2020; DOI: 10.1016/j.scitotenv.2020.140000.

[53] GOLD (Global Initiative for Chronic Obstructive Lung Disease). Global strategy for the diagnosis management and prevention of COPD. 2016; http://www.goldcopd.org/.

[54] Kurmi OP, Semple S, Simkhada P, et al. COPD and chronic bronchitis risk of indoor air pollution from solid fuel: a systematic review and metaanalysis. Thorax. 2010;**65**:221-228

[55] Hu G, Zhou Y, Tian J, et al. Risk of COPD from exposure to biomass smoke: a metaanalysis. Chest. 2010;**138**:20-31

[56] Sana A, Somda SMA, Meda N, et al. Chronic obstructive pulmonary disease associated with biomass fuel use in women: a systematic review and meta-analysis. BMJ Open Resp Res. 2018. DOI: 10.1136/bmjresp-2017-000246

[57] Hansel NN, McCormack MC, Belli AJ, et al. In-home air pollution is linked to respiratory morbidity in former smokers with chronic obstructive pulmonary disease. American Journal of Respiratory and Critical Care Medicine. 2013;**187**:1085-1090

[58] Johnston FH, Purdie S, Jalaludin B, et al. Air pollution events from forest fires and emergency department attendances in Sydney, Australia 1996-2007: A case-crossover analysis. Environmental Health. 2014;**13**:105

[59] Zhou Y, Zou Y, Li X, et al. Lung function and incidence of chronic obstructive pulmonary disease after improved cooking fuels and kitchen ventilation: a 9-year prospective cohort study. PLoS Medicine. 2014;**11**:e1001621

[60] Thakur M, Nuyts PAW, Boudewijns EA, et al. Impact of improved cookstoves on women's and child health in low and middle income countries: a systemic review and meta-analysis. Thorax. 2018;**73**:1026-1040

[61] Asher I, Pearce N. Global burden of asthma among children. The International Journal of Tuberculosis and Lung Disease. 2014;**18**:1269-1278

[62] Barry AC, Mannino DM, Hopenhayn C, et al. Exposure to indoor biomass fuel pollutants and asthma prevalence in South eastern Kentucky: results from the burden of Lung Disease (BOLD) study. The Journal of Asthma. 2010;**47**:735-741

[63] Oluwole O, Arinola GO, Huo D, et al. Biomass fuel exposure and asthma symptoms among rural school children in Nigeria. The Journal of Asthma. 2017;**54**:347-356

[64] Oluwole O, Arinola GO, Huo D, et al. Household biomass fuel use, asthma symptoms severity, and asthma under diagnosis in rural schoolchildren in Nigeria: a cross-sectional observational study. BMC Pulmonary Medicine. 2017;**17**:3

[65] Ayuk AC, Ramjith J, Zar HJ, et al. Environmental risk factors for asthma in 13-14 year old African children. Paediatric pulmonology. 2018. DOI: 10.1002/ppul.24162

[66] Thacher JD, Emmelin A, Aboi JK, et al. Biomass fuel use and the risk of asthma in Nigeria children. Respiratory Medicine. 2013;**107**:1845-1851

[67] Gaviola C, Miele CH, Wise RA, et al. Urbanisation but not biomass

fuel smoke exposure is associated with asthma prevalence in four resource-limited settings. Thorax. 2016;**71**:154-160

[68] Noonan CW, Semmens EO, Smith P, et al. Randomized trial of interventions to improve childhood asthma in homes with wood burning stoves. Environmental Health Perspectives. 2017;**125**:097010

[69] Shah K, Kunal S, Gothi R. Bronchial anthracofibrosis: the spectrum of radiological appearances. Indian J Radiol Imaging. 2018;**28**:333-341

[70] IHME (Institute for Health Metrics and Evaluation). 2016. "GBD Compare Data Visualization." IHME, University of Washington, Seattle. http://vizhub.healthdata.org/gbd-comparevizhub.healthdata.org/gbd-compare.

[71] IARC. Household use of solid fuels and high temperature frying. IARC Monographs on the evaluation of carcinogenic risks to humans. 2010;95

[72] Bruce N, Dherani M, Liu R, et al. does household use of biomass fuel cause lung cancer? A systemic review and evaluation of the evidence for the GBD 2010 study. Thorax. 2015;**70**:433-441

[73] Raspanti GA, Hashibe M, Siwakoti B, et al. Household air pollution and lung cancer risk among never-smokers in Nepal. Environmental Research. 2016;**147**:141-145

[74] Ferlay J, Soerjomataram I, Ervik M, et al. GLOBOCAN 2012 v1.0, Cancer Incidence and Mortality Worldwide: IARC CancerBase No. 11 [Internet]. Lyon, France: International Agency for Research on Cancer; 2013.http://globocan.iarc.fr.

[75] Okello S, Akello SJ, Dwomoh E, et al. Biomass fuel as a risk factor f oesophageal squamous cell carcinoma: a systemic review and meta-analysis. Environmental Health. 2019;**18**:60. DOI: 10.1186/s12940-019-0596-0

[76] Josyula S, Lin J, Xue X, et al. Household air pollution and cancers other than lung: a metaanalysis. Environmental Health. 2015;**14**:24. DOI: 10.1186/s12940-015-0001-3

[77] Smith KR, Pillarisetti A. Household Air Pollution from Solid Cook fuels and its Effects on Health. In: Mock CN, Nugent R, Kobusingye O, et al, editors. Injury prevention and Environmental Health. 3rd edition. Washington (DC).

[78] Kulkarni H, Narlawar UE, Sukksohale ND, et al. Biomass fuel use and risk of cataract: systemic review and meta-analysis. British J of medicine and medical research. 2014;**4**:382-394

[79] Yusuf F, Joseph P, Rangarajan S, et al. Modifiable risk factors, cardiovascular disease, and mortality in 155, 722 individuals from 21 high-income, middle-income, and low-income countries (PURE): a prospective cohort study. Lancet. 2020;7:795-808

[80] WHO. www.who.int/news-room/fact-sheets/detals/household-air-pollution-and-health.

[81] Fatmi Z, Coggon D. Coronary heart disease and household air pollution from use of solid fuel: a systemic review. British Medical Bulletin. 2016;**118**:95-113

[82] Arku RE, Ezzati M, Baumgartner J, et al. Elevated blood pressure and household solid fuel use in premenopausal women: Analysis of 12 Demographic and Health Surveys (DHS) from 10 countries. Environmental Research. 2018;**160**:499-505

[83] Ofori SN, Fobil JN, Odia OJ. Household biomass fuel use, blood

pressure and carotid intima media thickness; a cross sectional study of rural dwelling women in Southern Nigeria. Environ Pollut 2018;242(Pt A):390-397.

[84] Mocumbi AO, Stewart S, Patel S, et al. Cardiovascular Effects of Indoor Air Pollution from Solid Fuel: Relevance to Sub-Saharan Africa. Curr Environ Health Rep. 2019;**6**(3):116-126

[85] Pearson JF, Bachireddy C, Shyamprasad S, et al. Association between fine particulate matter and diabetes prevalence in the U.S. Diabetes Care. 2010;**33**:2196-2201

[86] Amegah AK, Quansah R, Jaakkola JJK. Household Air Pollution from Solid Fuel Use and Risk of Adverse Pregnancy Outcomes: A Systematic Review and Meta-Analysis of the Empirical Evidence. PLoS One. 2014;**9**:e113920. DOI: https://doi.org/10.1371/journal.pone.0113920

[87] Thompson LM, Bruce N, Eskenazi B, et al. Impact of Reduced Maternal Exposures to Wood Smoke from an Introduced Chimney Stove on Newborn Birth Weight in Rural Guatemala. Environmental Health Perspectives. 2011;**119**:1489-1494

[88] Mukherjee S, Siddique S, Chakraborty S, et al. Adverse reproductive health outcomes in pre-menopausal Indian women chronically exposed to biomass smoke. Journal of Public Health. 2015;**23**:363-372

[89] Agrawal S, Yamamoto S. Effect of indoor air pollution from biomass and solid fuel combustion on symptoms of preeclampsia/eclampsia in Indian women. Indoor Air. 2015;**25**(3): 341-352

[90] Yu H, Yin Y, Zhang, J. et al. The impact of particulate matter 2.5 on the risk of preeclampsia: an updated systematic review and meta-analysis. Environ Sci Pollut Res 2020. doi.org/10.1007/s11356-020-10112-8

[91] Weber E, Adu-Bonsaffoh K, Vermeulen R, et al. Household fuel use and adverse pregnancy outcomes in a Ghanaian cohort study. Reproductive Health. 2020;**17**(1):29. DOI: 10.1186/s12978-020-0878-3

[92] Hahad O, Lelieveld J, Birklein F, et al. Ambient Air Pollution Increases the Risk of Cerebrovascular and Neuropsychiatric Disorders through Induction of Inflammation and Oxidative Stress. International Journal of Molecular Sciences. 2020;**21**(12):4306. DOI: 10.3390/ijms21124306

[93] Thurston GD, Kipen H, Annesi-Maesano I, et al. A joint ERS/ATS policy statement: what constitutes an adverse health effect of air pollution? An analytical framework. The European Respiratory Journal. 2017;**49**(1):1600419

[94] Kioumourtzoglou MA, Schwartz JD, Weisskopf MG, et al. Long-term PM2.5 Exposure and Neurological Hospital Admissions in the Northeastern United States. Environmental Health Perspectives. 2016;**124**(1):23-29

[95] Perera FP, Li ZG, Whyatt R, et al. Prenatal airborne polycyclic aromatic hydrocarbon exposure and child IQ at age 5 years. Pediatrics. 2009;**124**:E195-E202

[96] Guxens M, Aguilera I, Ballester F, et al. Prenatal exposure to residential air pollution and infant mental development: modulation by antioxidants and detoxification factors. Environmental Health Perspectives. 2012;**120**:144-149

[97] Lam J, Sutton P, Kalkbrenner A, Windham G, et al. A Systematic Review and Meta-Analysis of Multiple Airborne Pollutants and Autism

Spectrum Disorder. PLoS One. 2016;**11**(9):e0161851

[98] Schraufnagel DE, Balmes JR, Cowl CT, et al. Air Pollution and Noncommunicable Diseases A Review by the Forum of International Respiratory Societies' Environmental Committee, Part 2: Air Pollution and Organ Systems. Chest. 2019;**155**(2):417-426

[99] Chan KH, Bennett DA, Kurmi OP, Yang L, Chen Y, Lv J, Guo Y, Bian Z, Yu C, Chen X, Dong C, Li L, Chen Z, Lam KBH; China Kadoorie Biobank Study Group. Solid fuels for cooking and tobacco use and risk of major chronic liver disease mortality: a prospective cohort study of 0.5 million Chinese adults. International Journal of Epidemiology 2020 ;49(1):45-55.

[100] Araviiskaia E, Berardesca E, Bieber T, et al. The impact of airborne pollution on skin. Journal of the European Academy of Dermatology and Venereology. 2019;**33**(8):1496-1505

[101] Nguyen VH. Environmental Air Pollution and the Risk of Osteoporosis and Bone Fractures. Journal of Preventive Medicine and Public Health. 2018;**51**(4):215-216

[102] Zhao CN, Xu Z, Wu GC, et al. Emerging role of air pollution in autoimmune diseases. Autoimmunity Reviews. 2019;**18**(6):607-614

Section 2

Environmental Emissions Monitoring and Mitigation

Chapter 3

Importance of Air Quality Networks in Controlling Exposure to Air Pollution

David Galán Madruga

Abstract

An air quality monitoring network (AQMN) is a basic piece of environmental management due to that it satisfies the major role in monitoring of environment emissions, in special relevance to target air pollutants. An adequate installation would lead to support high efficiency of the network. Therefore, AQMN pre-layout should be considered as an essential factor in regarding with the location of fixed measurement stations within AQMN, as the minimum number of sampling points. Nevertheless, once AQMN has been already installed, and given that the spatial air pollutants pattern can vary along time, an assessment of the AQMN design would be addressed in order to identify the presence of potential redundant fixed monitoring stations. This approach would let to improve the AQMN performance, reduce maintenance costs of the network and consolidate the investment on those more efficient fixed stations. The chapter includes aspects relative to air pollutants measured by networks, their representativeness, limitations, importance, and the future needs. It ponders the need of re-assessment of the AQMN layout for assuring (i) a right evaluation of the human being exposure to atmospheric pollutants and controlling the environmental emissions into the atmosphere and (ii) an adequate performance of the network along time.

Keywords: environmental emissions, air quality, fixed monitoring stations, design and performance

1. Introduction

While technological advances have generated an improvement in the human being's life quality, they have also contributed to the emergence of associated issues, such as exponential industrial grown and increase of transportation networks, due to a fast growing population and its centralization into urban centers, mainly. As a consequence, the rise of the pollutant emissions toward the environmental compartments has been framed as a Public Health concern. Therefore, the impact of environmental emissions on the climate and the environment is an ultimate subject, both to local, regional as global level. Particularly, an increase of environmental well-being would bring in greater the quality of life, due to the exchange internal-external between the human being and the environment.

Environmental emissions play a key role in the release of pollutants on air, water and soil matrix, which is relevant because drives a decline of biodiversity [1]. In this sense, deforestation, water pollution, acid rain or endangered animals are factors linked to likely environmental repercussions [2, 3]. In the health framework, numerous epidemiological studies associate the presence of pollutants in the environment and harmful effects on human being health [4, 5].

Taking into account the interaction between human being and environmental, the human well-being is a factor tied to the presence of clean air, otherwise, the emergence of harmful effects could drive to devastating implications on human health. According to the 68th World Health Assembly (see [6]), each year, a total of 4.3 and 3.7 million deaths result from exposure to indoor and outdoor pollutants, respectively.

Among different environmental compartments, this chapter will focus on atmospheric matrix, given that atmospheric pollution is considered the major environmental risk to human health worldwide. Atmospheric pollution result from the release of a damaging chemical or material into the atmosphere and it encompasses a wide variety of pollutants, either organic or inorganic compounds. Once air pollutants are released into the atmosphere, those ones can be exhibited both gaseous phase as solid and liquid particles suspended in the air (particulate material, PM) [7].

The occurrence of pollutants in the atmosphere depends on emissions sources. Although the atmospheric pollution is considered as a global character issue, the highest levels of air pollutants have been monitored in the developing countries, as a consequence of the industrial growth. A more detailed analysis would pointed to large cities [8, 9], where environmental emissions to atmospheric level could come from several types of sources of pollution, such as industrial developments implying combustion processes, vehicular emissions and domestic-heating [10].

Environmental emissions have a direct effect on the outdoor ambient pollution, as well as on indoor air quality, given that outdoor emission sources are responsible for the presence of air pollutants at indoor environments [11], due to the gases and particles infiltration.

Based on previously mentioned, atmospheric pollution monitoring is of a fundamental importance in Public Health [12] in order to (i) control the human being exposure to air pollutants [13] and (ii) support the decision making on environmental management, in particular air quality management [14]. So, for example, an adequate management of major dominant emission sources in urban environments, as it can be to limit the road transport and more restricted industrial emissions, would result in lower levels of pollution into the atmosphere.

European Union develops Air Quality Directives [15] for setting air quality objectives in order to reduce potential harmful effects of air pollutants on human health and environment, establishing limit and target values for criteria of air pollutants. Air quality assessment is a responsibility of each Member States within their territory. These ones have the obligation for maintaining an air quality good, or improve it, and they should assume action in order to comply with the limit values and critical levels and, where possible, to reach the target values and long-term objectives. For that, Member States establish air quality monitoring networks (AQMN) in their territories for verifying compliance with those air quality objectives.

Therefore, AQMN performs an essential function within Public Health framework, monitoring environment emissions and controlling exposure in order to take care of human being health.

2. Criteria of air pollutants measured at AQMN

Criteria of air pollutants are those atmospheric pollutants generally monitored by AQMNs. They usually measure the next legislated criteria of air pollutants (see **Table 1**): sulfur dioxide, nitrogen oxides (monoxide and dioxide nitrogen), benzene, carbon monoxide. Ozone and atmospheric particles (PM_{10} particles, with an aerodynamic diameter of 10 µm or less, and $PM_{2.5}$, aerodynamic diameters ≤2.5 µm) [16, 17].

The measurements recorded by AQMNs must be able to provide traceability, in order to compare air quality data among all Member States, for which those measures must be monitored using common measurement methods. For that, Member States apply normalized reference measurement methods (see **Table 2**).

The previously mentioned methods are cataloged as automatic method, nevertheless, the AQMNs dispose manual methods for determining the chemical composition of atmospheric particles (PM_{10} and $PM_{2.5}$), mainly for heavy metals and polycyclic aromatic hydrocarbons. While the samples are collected by manual equipment installed at AQMN, their composition is analyzed in the laboratory.

Pollutant	Quantitative value	Concept	Averaging period
Sulfur dioxide (gas)	350 µg/m^3 125 µg/m^3	Limit value Limit value	1 hour 1 day
Nitrogen dioxide (gas)	200 µg/m^3 40 µg/m^3	Limit value Limit value	1 hour A calendar year
Benzene (gas)	5 µg/m^3	Limit value	A calendar year
Carbon monoxide (gas)	10 mg/m^3	Limit value	Maximum daily 8-hour mean
Ozone (gas)	120 µg/m^3	Target value (January 1, 2010)	Maximum daily 8-hour mean
PM_{10} (particles)	50 µg/m^3 40 µg/m^3	Limit value Limit value	1 day A calendar year
$PM_{2.5}$ (particles)	25 µg/m^3 20 µg/m^3	Target value (January 1, 2020) Limit value	A calendar year A calendar year

Table 1.
Limit and target value for the protection of human health [15].

Pollutant	European standard	Measurement method
Sulfur dioxide	EN 14212:2012 [18]	Ultraviolet fluorescence
Monoxide and nitrogen dioxide	EN 14211:2012 [19]	Chemiluminescence
Benzene	EN 14662:2005 [20]	Automated pumped sampling with in situ gas chromatography
Carbon monoxide	EN 14626:2012 [21]	Non-dispersive infrared spectroscopy
Ozone	EN 14625:2005 [22]	Ultraviolet photometry
PM_{10} and $PM_{2.5}$	EN 12341:2015 [23]	Gravimetric method

Table 2.
Reference measurement methods for measuring criteria of air pollutants.

The occurrence of criteria of air pollutants into the atmosphere measured at fixed stations within AQMN is dependent on several factors, such as meteorological features and sources of pollution.

3. Types of air pollutant emission sources

The comprehension of emission sources and knowledge on pollution levels reached in the air matrix could be useful tool for understanding the spatial and temporal distribution of air pollutants, which would provide an overview picture about human exposure to environmental emissions coming from different sources of contamination.

In function of their origin, it is necessary to distinct between anthropogenic and natural sources. Broadly, the first ones are sources that release mixtures of pollutants come from transport, power generation, industrial activity, biomass burning, and domestic heating, mainly in urban environments [24–26] while volcanic eruptions, plant emission and oceans are tied to natural sources. Nevertheless, in terms of released pollutant, the sources can be defined as primary and secondary. Primary emission sources result from the direct emissions from an air pollution source, while secondary emission sources result from the formation of a pollutant in the atmosphere from the chemical reaction between theirs precursors, which are emitted from air pollution primaries sources, and the meteorological variables. Finally, once pollutants are released, either from primary or secondary sources, the pollutants can be deposited on the Earth's terrestrial or aquatic surfaces, followed by re-emission to the atmosphere; in this case the sources are named as re-emission sources [27].

While the identification of emissions sources is a fundamental factor in order to carry out the distribution of the fixed monitoring stations within an AQMN, other elements also perform a primordial role, such as population density, peculiar features of target territory, amplitude of geographic area to be controlled as further meteorological variables.

4. Air quality monitoring network

4.1 Concept

AQMN is an essential element within environmental management, in special emphasis to air quality management. It is consisted of fixed monitoring stations for measuring air pollutants (see **Figure 1**). Although the total number of stations depends on several factors, according to Section 3, these ones should be attributed conveniently in the domain of interest for providing suitable air pollutant information and estimating the exposure of the ambient pollution on human being of righter way. So, one of the keys in the AQMN layout is the distribution of monitoring stations as well as the determination of a sufficient and confident number of sampling points for carrying out the air quality measurements. These features are associated with the network management, which should focus on reducing the fixed stations within the AQMN to a reliable and non-redundant number. So, the network would not duplicate information on air pollution.

Given that the assessment of air quality in Member States is approached by means of the data generated by AQMN, those ones divide their territories at zones and agglomerations in order to reach that objective. Generally, air quality must be assessed in all zones and agglomerations by means of one or more fixed stations.

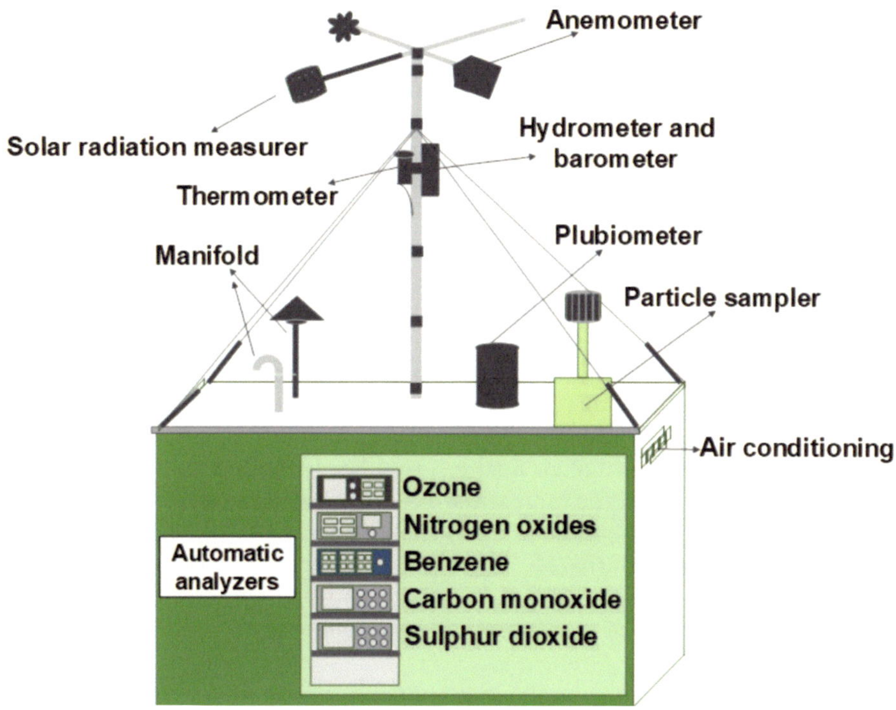

Figure 1.
Basic setup of a fixed monitoring station.

The number of zones can vary in function of geographical location, distribution of emission sources and meteorology, although the final number of those ones must provide an adequate representation of the territory heterogeneity.

Current European legislation [15] lays down criteria for siting fixed stations within an AQMN, pointing a wide number of considerations regards at macroscale siting of sampling points in order to protect the human health, vegetation and natural ecosystems. Similarly, in terms of microscale, the legislation set criteria relative to air flow no-restriction around the inlet of sampling point, its height regards to ground level (between 1.5 and 4 m) and distance regards to the edge of major junctions (at least 25 m) and the kerbside (no more than 10 m).

Regarding to minimum number of fixed monitoring stations, the European legislation set criteria in those zones and agglomerations where fixed measurement is the sole source of information for evaluating compliance with limit values for the protection of human health and alert thresholds. This criterion is based on zone inhabitant number and measured pollutants.

Fixed stations included into AQMN can be sorted in function of several typologies. So, in terms of area where is located it can be named:

- Urban stations: Those ones located at zones with presence of buildings of continued way.
- Suburban stations: Those ones located at zones with presence of buildings of continued way but separated by no-buildings areas, such as lakes, forests and agricultural land.
- Rural stations: Those ones located at zones not included within the previous two categories.

In terms of major dominant emission source:

- Traffic stations: Those ones which contamination levels are mainly appointed to emission sources coming from vehicles.
- Industrial stations: Those ones which contamination levels are majorly dependent on industrial activities.
- Background stations: Those ones which contamination levels cannot be directly attributed to any dominant emission source.

4.2 Representativeness of fixed monitoring stations within AQMN

Regardless the function exhibited by fixed measurement stations included into an AQMN, as the assessment of air quality, cross-border pollution, spatial-temporal trends or exposure studies, the representativeness of each station should be considered as a primordial reflection. The efficiency degree of the fixed stations into AQMN can be assessed in terms of:

4.2.1 Representation degree of any station within its zone or agglomeration

Given that one target zone can be represented by one or more fixed stations, it is relevant to know the spatial representativeness of each station in order to evaluate whether air quality monitored by those ones can or not be extrapolated to all zone. In this sense, in order to provide an overview regards to atmospheric pollution within zone, the passive methodology simultaneously samples a large number of sampling points, which supplies opportune information on spatial pollution in the researched zone [28]. This approach lets to compare air quality data measured by AQMN vs. those ones monitored by passive methods, thereby confirming or not the station representativeness within target zone.

4.2.2 Whole contribution of any station regarding environmental pollution data recorded by AQMN

Spatial representativeness of the information provided by AQMN is dependent on type of station, in terms of spatial scale, and the pollutant.

Broadly, representativeness of a fixed station can be defined as the variability of the target pollutant concentrations around sampling point, while others authors enlarged the definition to the radius of a circular area where the concentration can vary up to ±20%, as maximum value [29].

The AQMN performance does not depend on number of fixed measurement stations, given that the presence of redundant stations could result in existence of non-efficient fixed stations. This means that potential emission sources close to those stations could have a strong probability of similitude. For this reason, the representativeness of each station within an AQMN should be properly studied.

Although a final agreement regarding on procedure for assessing the efficiency of fixed stations have not been reached, this subject have been widely studied in the scientific literature. Recent studies have reported on approaches in order to test the AQMN performance. They are based on the use of several combined chemometric techniques for reached the aforementioned objective. Some authors have used the analysis of correlation for revealing the existence of redundant fixed stations, although this method does not identify efficient stations. For that, they apply the principal component analysis technique, which appoints a new set of linearly uncorrelated stations [29]. Other studies have expanded the number of chemometric techniques, combining correlation analysis, principal component analysis, assignment method, clustering analysis and correspondence analysis [30].

A significate consideration would drive to know the contribution degree of each source of pollution in regarding with air pollutants levels reached into the atmosphere. Some authors have solved this subject using a combination of techniques, combined principal component analysis and multiple linear regression.

Nevertheless, the practical application of the mentioned approaches were not tested along a period did not include in the development of these approaches. Similarly, they did not assess spatial information percentage that is lost when redesigning the AQMN due to the removal redundant fixed measurement stations.

Therefore, while different approaches have been developed to estimate the area of spatial representativeness of monitoring stations, a unique robust methodology to assess the representativeness of in-situ measurements has not yet been agreed.

Spatial representativeness is required for distinct actions:

- Station classification and network design [31],
- Air quality and exposure assessment, for example, to estimate the air quality standards exceedance areas and to quantify the population exposure to the air pollution [32],
- Model validation and data assimilation [33].

A lack of information in regarding with the AQMN performance, based on the representativeness of their fixed stations, would support their potential limitations.

4.3 Potential limitations of AQMN

4.3.1 Relative to the design

Given that air quality in any zone, either local, regional or global, is dependent on a wide number of factors such emission sources (transport network and industries) and meteorological features, the assessment of atmospheric pollution is a hard assignment, and due to these factors it is specific for each zone. Therefore, it is possible that the spatial information on environmental pollution reported by AQMNs is not representative of the target zone.

This limitation could influence on AQMN efficiency, if its function is framed within activity of informing population of levels which they are exposed to. This fact is relevant because numerous epidemiological studies use air quality data recorded by AQMNs, in order to associate air pollutant levels with damaging effects or hospital admissions. Nevertheless, the pollution data measured by AQMNs along study time is not equivalent to the daily concentrations which the human being is exposed to. So, the reached conclusions could exhibit a limited scope.

4.3.2 Relative to current European legislation

While European Legislation clearly establish criteria, on the one hand, for siting potential fixed measurement stations and, on the other hand, setting minimum number of those ones, criteria for identifying the more representative fixed sampling points within AQMN is not considered. This fact is fundamental in order to optimize the AQMN performance.

4.3.3 Relative to the development of specific procedure for evaluating the representativeness of fixed stations

FARIMODE study [34] reported on collected information coming from questionnaire to get technical information concerning the methodologies used to estimate the representativeness of air quality monitoring stations. The questionnaire was answered by 22 workgroups from 14 different countries providing information on 25 methodologies.

Major methodological limitations were appointed to input data availability (9 answers), expert or local knowledge (1), modeling domain (1), modeling uncertainties (6), input data uncertainties (10), temporal-spatial resolution (7), directive metrics (1), computational resources (4), pollutants (2), definition of parameters of methodology (3), coverage of station network (1) and no limitation (2). Within this study, a relevant conclusion was the possibility for examining if the similarities or discrepancies between the representativeness estimates are more or less significant according to the concentration levels measured by target station.

4.4 Importance of AQMNs

Consequently to previously mentioned, an AQMN play a paramount lead in the evaluation of the air quality [35], in order to:

- Inform to the population in regarding with pollution levels which they are exposed to [30],
- Know spatial-temporal pattern of air pollutants (see **Figure 1**),
- Identify predominant emission sources [36],
- Support the development of monitoring strategies [37] and.
- Assist to authorities in decisions making.

Besides, in the case of Member States, those ones must report to European Commission on recorded pollution data, which lets to evaluate the cross-border pollution and model the spatial-temporal air pollution pattern, among others applications.

AQMN links important subjects framed into Public Health, such as sources of pollution, environmental emissions, outdoor air pollutant levels and human health. Therefore, AQMN proves a helpful implement for estimating risk associated with human being exposure to air pollutant levels occurred into the atmosphere.

4.5 Management of AQMNs

Within the European Union, Member States are liable for controlling and assuring data quality of fixed monitoring stations. Each one establishes the necessary number and the location of fixed measurement stations included in their AQMNs, in order to ensure an adequate air quality assessment in its territory and comply with air quality standards. Similarly, each Member State is responsible for managing their AQMNs, according to requirements set in current European Legislation, meaning, the used measurement methods should be those ones included in air quality standards (they have been mentioned in **Table 2**). Similarly, they should guarantee a proper maintenance of those measure devices employed for monitoring atmospheric pollutants in the outdoor ambient air. For that, the air quality standards [18–23] set several basic qualifications regards to the measure devices and their management:

- Components of sampling system: (i) Sampling line: standards indicate the frequency of clear or, if necessary, its change, (ii) Particle filters: standards indicate where should site and (iii) Sampling pump: standards set the sampling flux required for working properly.

- Equipment requirements. They depend on target atmospheric pollutant. The components of the devices used for measuring atmospheric pollutants are described in the Air Quality Standards. The next devices can be differenced:

Continuous devices: The air pollutant levels are continuously measured using automatic analyzers.

Integrated devices: Levels of the target air pollutants are measured by manual or automated methods and the data is registered hourly or daily.

Static devices: Levels of the target air pollutants are estimated by using qualitative measurement devices from weekly or monthly exposure.

- Maintenance operations: The Air Quality Standards determine those necessary actions in order to test if an equipment is working within specifications marked by the manufacturer. For that aim, technical aspects such as verification of zero, the higher concentration level and lack of fit, among other should be checked. All these tests should provide satisfactory results, complying with the criteria set in the air quality standards.

- Equipment calibration: Standards exhibit the frequency of calibration for each criteria of air pollutant, as well as recommended concentration/s, acceptation criteria, methodology, etc.

- Quality control and quality assurance: The execution of this subject assures that the uncertainty or dispersion associated with the measured values by AQMNs fall down criteria set by current European Legislation. For that, the compliance of the previous requirements should be reached.

More detailed information about this section can be found in air quality standards, which have been published by CEN/TC 624 Work Programme and can be acquired through European Committee for Standardization (https://www.cen.eu/Pages/default.aspx, accessed March 6, 2020).

4.6 Data measured by AQMNs

Given that the AQMNs management is a responsibility for the Member States, those ones should ensure valid air quality data. The data registered during maintenance, check and calibration processes should be not included within the air quality dataset, as well as the faulty data. The air quality standards establish requirements relative to the way for expressing the air quality data (number of decimals) and data capture (temporal coverage).

At the State level, Member States transfer air quality data to Europe Union. Those data can be compared given that, on the one hand, their measure was monitored using reference methods and, on the other hand, they complied with those QC/QA criteria set by air quality standards. As an example, the European

Monitoring and Evaluation Programme (EMEP) is the co-operative program for assessing long range transmission of atmospheric pollutants over Europe. Member States have an air quality network in order to monitor background levels of air pollutants. Information relative to this subject can be found in https://www.emep.int/ (accessed March 6, 2020), where emission data, measurement data and modeling results for air pollutants are available through an open's database. Similarly, air quality data monitored by Member States is reported by European Environment Agency (https://www.eea.europa.eu/data-and-maps/explore-interactive-maps/up-to-date-air-quality-data, accessed March 9, 2020) by means of interactive maps and reports. Other website providing real-time air pollution data by interactive maps in Europe and other countries over the world can be visited at https://aqicn.org/map/europe/ (accessed March 9, 2020).

4.7 Potential suggestions for improving the AQMN management

Given that the AQMNs have a large historical series of ambient air data for target air pollutants [38] (sulfur dioxide, nitrogen monoxide and dioxide, benzene, carbon monoxide, ozone and atmospheric particles), a count of the number of times that the measurements exceeded the limit and target value established by the European Legislation would help to identify those air pollutants who should be monitored.

As a consequence, the measurement of those air pollutants which do not exceeded the limit values could be reduced in terms of number of fixed monitoring stations, which would give to reinvest those economic resources towards the monitor of other pollutants, e.g. benzene, given that the measurements of this last pollutant are very limited within AQMNs and, according to European Legislation, its measure is mandatory.

Based on the role of AQMNs within environmental emission control, and given that their spatial monitoring coverture is limited [39], nowadays, new wireless low cost sensors are available in order to assess pollution levels in ambient [40] and indoor air [41], by simultaneous monitoring in an elevated number of sampling points,

5. Possible future trends

At European level, although the application of measurement methods for monitoring air pollutants in AQMNs is normalized by Air Quality Standards, providing traceability to air pollution data among Member States, a harmonized technique for estimating the representativeness of fixed monitoring stations have not been defined.

In the future, the major requirement in regards to AQMN design should point towards the development of a particular methodology for evaluating representativeness of fixed monitoring sampling points within a network.

On the one hand, this methodology should offer evaluation criteria which would assure an adequate estimation of the representativeness and, on the other hand, they should be common and similar to all Member States.

The implementation of this reflection would result in a significant benefit for population, given that an optimization regards to location of fixed stations, and by extension on AQMN performance, it would aid to control the human exposure to atmospheric pollution in a more precise way, supporting a more realistic estimate of human health risk.

6. Conclusions

The binomial between environmental emissions and human exposure leads to Public Health concerns. In particular, emissions towards ambient air are considered the higher environmental risk. In order to control those issues, AQMNs play a paramount role for controlling air pollution in order to evaluate the compliance with those air quality objectives set by Air Quality European Standards and assist to authorities in decisions making. They consist of fixed monitoring stations and measure several criteria of air pollutants. Although each fixed station should be representative of an around area, the spatial coverture of AQMNs is very limited, due to the restricted number of sampling points as a consequence of the large investment need for setting up an AQMN. Current European legislation lays down criteria for supporting location and minimum number of the fixed measurement stations within AQMN. There are numerous websites exhibiting air quality data (at local, regional and global level), by means of reports, interactive maps or time-real data.

In order to support the AQMN management, a study regards to number of times that air pollutant measures have exceeded the criteria of air quality set in European Legislation should be addressed, for identifying the pollutants which should be measured.

A relevant subject of an AQMN would point to its layout. Although, the legislation does not set methods for evaluating representativeness of the fixed measurement points or requirements for refereeing representativeness degree, this one should be tested over the time, given that new emission air pollutant sources can be emerged, which would directly affect to the AQMN performance.

Therefore, the deployment of a harmonized methodological framework is required, which allows to establish a comprehensive and comparative evaluation of the AQMN efficacy, by evaluating the representativeness of fixed monitoring stations.

This methodology should be assisted by scientists, AQMN's managers and technicians and experts of air quality and it should lay down the concrete type of method to use, either passive methodology, modeling, series of historical data, a combination of them or other methods.

The development of this harmonized methodological would help to the reporting of spatial representativeness by the Member States to Commission European by means of a common approach.

Conflict of interest

The author declares no conflict of interest. This work does not have commercial purposes, only scientific ones.

Other declarations

This work did not receive any specific grant from funding agencies in the public, commercial, or not-for-profit sectors.

Author details

David Galán Madruga
Atmospheric Pollution Area, National Center for Environment Health, Carlos III Health Institute, Madrid, Spain

*Address all correspondence to: david.galan@isciii.es

References

[1] Avigliano E, Rosso JJ, Lijtmaer D, Ondarza P, Piacentini L, Izquierdo M, et al. Biodiversity and threats in non-protected areas: A multidisciplinary and multi-taxa approach focused on the Atlantic Forest. Heliyon. 2019;**5**(8):e02292

[2] Chi Y, Yang P, Ren S, Ma N, Yang J, Xu Y. Effects of fertilizer types and water quality on carbon dioxide emissions from soil in wheat-maize rotations. Science of the Total Environment. 2020;**698**:134010

[3] Liu Z, Li D, Zhang J, Saleem M, Zhang Y, Ma R, et al. Effect of simulated acid rain on soil CO_2, CH_4 and N_2O emissions and microbial communities in an agricultural soil. Geoderma. 2020;**366**:114222

[4] Kim E, Park H, Hong Y-C, Ha M, Kim Y, Kim B-N, et al. Prenatal exposure to PM_{10} and NO_2 and children's neurodevelopment from birth to 24 months of age: Mothers and Children's Environmental Health (MOCEH) study. Science of the Total Environment. 2014;**481**:439-445

[5] Zhang G, Jiang F, Chen Q, Yang H, Zhou N, Sun L, et al. Associations of ambient air pollutant exposure with seminal plasma MDA, sperm mtDNA copy number, and mtDNA integrity. Environment International. 2020;**136**:105483

[6] World Health Organization. Sixty-eighth World Health Assembly. A68/18. In: Health and the Environment: Addressing the Health Impact of Air Pollution. World Health Organization (WHO); 2015. Available from: https://apps.who.int/gb/ebwha/pdf_files/WHA68/A68_18-en.pdf

[7] Wang G, Yu J, Su Y, Shi G. Distribution and regeneration of hydroxyl free radicals in gaseous and particulate phases of pollutants in near-ground ambient air. Science of the Total Environment. 2019;**683**:221-230

[8] Han C, Liu R, Luo H, Li G, Ma S, Chen J, et al. Pollution profiles of volatile organic compounds from different urban functional areas in Guangzhou China based on GC/MS and PTR-TOF-MS: Atmospheric environmental implications. Atmospheric Environment. 2019;**214**:116843

[9] Abbass RA, Kumar P, El-Gendy A. Car users exposure to particulate matter and gaseous air pollutants in megacity Cairo. Sustainable Cities and Society. 2020;**56**:102090

[10] González CM, Gómez CD, Rojas NY, Acevedo H, Aristizábal BH. Relative impact of on-road vehicular and point-source industrial emissions of air pollutants in a medium-sized Andean city. Atmospheric Environment. 2017;**152**:279-289

[11] Madruga DG, Ubeda RM, Terroba JM, dos Santos SG, García-Cambero JP. Particle-associated polycyclic aromatic hydrocarbons in a representative urban location (indoor-outdoor) from South Europe: Assessment of potential sources and cancer risk to humans. Indoor Air. 2019;**29**(5):817-827

[12] Omrani H, Omrani B, Parmentier B, Helbich M. Spatio-temporal data on the air pollutant nitrogen dioxide derived from Sentinel satellite for France. Data in Brief. 2020;**28**:105089

[13] Xing Y, Brimblecombe P. Urban park layout and exposure to traffic-derived air pollutants. Landscape and Urban Planning. 2020;**194**:103682

[14] McDonald F, Horwell CJ, Wecker R, Dominelli L, Loh M, Kamanyire R, et al. Facemask use for community protection from air pollution disasters: An ethical overview and framework to guide agency decision making. International Journal of Disaster Risk Reduction. 2020;**43**:101376

[15] Directive 2008/50/EC of the European Parliament and of the Council on 21 May 2008 on ambient air quality and cleaner air for Europe. 2008. Available from: https://eur-lex.europa.eu/legal-content/EN/TXT/PDF/?uri=CELEX:32008L0050&from=EN

[16] Yang J, Kang S, Ji Z, Yin X, Tripathee L. Investigating air pollutant concentrations, impact factors, and emission control strategies in western China by using a regional climate-chemistry model. Chemosphere. 2020;**246**:125767

[17] Loomis D, Grosse Y, Lauby-Secretan B, Ghissassi FE, Bouvard V, Benbrahim-Tallaa L, et al. The carcinogenicity of outdoor air pollution. The Lancet Oncology. 2013;**14**(13):1262-1263

[18] EN 14212:2012. Ambient air. Standard method for the measurement of the concentration of sulphur dioxide by ultraviolet fluorescence. 2012

[19] EN 14211:2012. Ambient air. Standard method for the measurement of the concentration of nitrogen dioxide and nitrogen monoxide by chemiluminescence. 2012

[20] EN 14662:2005. Ambient air quality. Standard method for measurement of benzene concentrations. pumped sampling followed by thermal desorption and gas chromatography. 2005

[21] EN 14626:2012. Ambient air. Standard method for the measurement of the concentration of carbon monoxide by non-dispersive infrared spectroscopy. 2012

[22] EN 14625:2012. Ambient air quality. Standard method for the measurement of the concentration of ozone by ultraviolet photometry. 2012

[23] EN 12341:2014. Ambient air. Standard gravimetric measurement method for the determination of the PM_{10} or $PM_{2.5}$ mass concentrations of suspended particulate matter. 2014

[24] Song J, Zhao C, Lin T, Li X, Prishchepov AV. Spatio-temporal patterns of traffic-related air pollutant emissions in different urban functional zones estimated by real-time video and deep learning technique. Journal of Cleaner Production. 2019;**238**:117881

[25] Zhou Y, Luo B, Li J, Hao Y, Yang W, Shi F, et al. Characteristics of six criteria air pollutants before, during, and after a severe air pollution episode caused by biomass burning in the southern Sichuan Basin China. Atmospheric Environment. 2019;**215**:116840

[26] Wang Y, Song J, Yang W, Dong L, Duan H. Unveiling the driving mechanism of air pollutant emissions from thermal power generation in China: A provincial-level spatiotemporal analysis. Resources, Conservation and Recycling. 2019;**151**:104447

[27] Outdoor Air Pollution. IARC Monographs on the Evaluation of Carcinogenic Risks to Humans. Vol. 109. International Agency for Research on Cancer; 2016. Available from: https://publications.iarc.fr/Book-And-Report-Series/Iarc-Monographs-On-The-Identification-Of-Carcinogenic-Hazards-To-Humans/Outdoor-Air-Pollution-2015

[28] Galán Madruga D, Fernández Patier R, Sintes Puertas MA, Romero García MD, Cristóbal LA. Characterization and

local emission sources for ammonia in an urban environment. Bulletin of Environmental Contamination and Toxicology. 2018;**100**(4):593-599

[29] Chow JC, Chen L-WA, Watson JG, Lowenthal DH, Magliano KA, Turkiewicz K, et al. $PM_{2.5}$ chemical composition and spatiotemporal variability during the California Regional PM_{10}/$PM_{2.5}$ Air Quality Study (CRPAQS): CRPAQS $PM_{2.5}$ spatiotemporal variability. Journal of Geophysical Research-Atmospheres. 2006;**111**(D10):1-17

[30] Zhao L, Xie Y, Wang J, Xu X. A performance assessment and adjustment program for air quality monitoring networks in Shanghai. Atmospheric Environment. 2015;**122**:382-392

[31] Henne S, Brunner D, Folini D, Solberg S, Klausen J, Buchmann B. Assessment of parameters describing representativeness of air quality in-situ measurement sites. Atmospheric Chemistry and Physics. 2010;**10**(8):3561-3581

[32] Malherbe L, Jimmink B, de Leeuw F, Schneider P, Ung A. Analysis of station classification and network design in EU28 (& other EEA) countries. EEA & ETC/ACM Working Paper. 2013

[33] Lefebvre W, Van Poppel M, Maiheu B, Janssen S, Dons E. Evaluation of the RIO-IFDM-street canyon model chain. Atmospheric Environment. 2013;77:325-337

[34] Martín F, Santiago JL, Kracht O, García L, Gerboles M. FAIRMODE spatial representativeness feasibility study. Report EUR 27385 EN. 2015

[35] Rosario L, Francesco SP. Analysis and characterization of the predominant pollutants in the Catania's air quality monitoring stations. Energy Procedia. 2016;**101**:337-344

[36] Pires JCM, Sousa SIV, Pereira MC, Alvim-Ferraz MCM, Martins FG. Management of air quality monitoring using principal component and cluster analysis—Part I: SO_2 and PM_{10}. Atmospheric Environment. 2008;**42**(6):1249-1260

[37] Kao J-J, Hsieh M-R. Utilizing multiobjective analysis to determine an air quality monitoring network in an industrial district. Atmospheric Environment. 2006;**40**(6):1092-1103

[38] Barrero MA, Orza JAG, Cabello M, Cantón L. Categorisation of air quality monitoring stations by evaluation of PM_{10} variability. Science of the Total Environment. 2015;**524**(525):225-236

[39] Munir S, Mayfield M, Coca D, Jubb SA. Structuring an integrated air quality monitoring network in large urban areas—Discussing the purpose, criteria and deployment strategy. Atmospheric Environment X. 2019;**2**:100027

[40] Molka-Danielsen J, Engelseth P, Wang H. Large scale integration of wireless sensor network technologies for air quality monitoring at a logistics shipping base. Journal of Industrial Information Integration. 2018;**10**:20-28

[41] Salman N, Kemp AH, Khan A, Noakes CJ. Real time wireless sensor network (WSN) based indoor air quality monitoring system. IFAC-Paper. 2019;**52**(24):324-327

Chapter 4

Industrial Air Emission Pollution: Potential Sources and Sustainable Mitigation

Rabia Munsif, Muhammad Zubair, Ayesha Aziz and Muhammad Nadeem Zafar

Abstract

Air of cities especially in the developing parts of the world is turning into a serious environmental interest. The air pollution is because of a complex interaction of dispersion and emission of toxic pollutants from manufactories. Air pollution caused due to the introduction of dust particles, gases, and smoke into the atmosphere exceeds the air quality levels. Air pollutants are the precursor of photochemical smog and acid rain that causes the asthmatic problems leading into serious illness of lung cancer, depletes the stratospheric ozone, and contributes in global warming. In the present industrial economy era, air pollution is an unavoidable product that cannot be completely removed but stern actions can reduce it. Pollution can be reduced through collective as well as individual contributions. There are multiple sources of air pollution, which are industries, fossil fuels, agro waste, and vehicular emissions. Industrial processes upgradation, energy efficiency, agricultural waste burning control, and fuel conversion are important aspects to reducing pollutants which create the industrial air pollution. Mitigations are necessary to reduce the threat of air pollution using the various applicable technologies like CO_2 sequestering, industrial energy efficiency, improving the combustion processes of the vehicular engines, and reducing the gas production from agriculture cultivations.

Keywords: environmental, pollution, industrial, emission, global warming

1. Introduction

A unique chemical wrapping that promotes life on glob and support numerous activities often referred to as air. Rapid industrialization is becoming serious concern for fresh air and healthy life [1–3]. Abundant discharge of industrial toxin making natural environment harmful, unstable, and uncomfortable for physical and also for biological environment and it leads to pollution by energy sources and chemical substances. Physical and biological environment are damage by the heat and pollutants in the air. These pollutants including vapors, aerosols, solid particles, toxic gases and smoke drive from industrial processes. Emission of air pollutants is also because of many human actions. List of six air pollutants presented by World health organization (WHO) which known as classic air pollutants in industrialized countries as nitrogen oxides (NO_x), sulfur dioxide (SO_2), carbon monoxide (CO), and suspended particulate matter [4]. A number of industrial sources are responsible for the emission of carbon monoxide along with, fuel-fired boilers, internal combustion gas boilers and gas stoves [5]. The quality of the combustion process is primary indicated by carbon dioxide. Emissions of CO_2, as a result of combustion of fuels, are creating consequences on environment [6]. For the industrial combustion system carbon dioxide was also examined a major greenhouse gas [7]. For the emission of carbon dioxide from any type of combustion source a prescribed national standard was present but it is important to check that carbon dioxide emission enter into air at steep rate. Oxides of nitrogen as nitrogen dioxide (NO_2) and nitric oxide (NO) produce from thermal power plants, vehicles, industrial process and, coal burning processes [8]. Oxides of nitrogen are produced by the reaction of free oxygen and nitrogen of air which achieved at high temperature during combustion process. Fuels, rich in sulfur contents produce sulfur dioxide (SO_2) gas when used for the energy. Bennett [9] reported sulfur dioxide lifetime is about 10 days in air. Industrial stacks emitting sulfur dioxide because fuels contain a standards higher concentration of sulfur. Generally, in Pakistan the electric power supply is not adequate and consistent for supporting employments; consequently, to overcome electric energy shortage all business sectors still extensively use their private generator (**Figure 1**).

These generators mostly installed next to their services or along the road which is not an appropriate location. Therefore, by importing exhaust gases into the air they create many complications to the people who are traveling on the roads and the resident. Smoke opacity of industrial gas is also a parameter which has considerable potential to enhance environmental air pollution by smoke particles emission. Industrial stack points were also analyzed with reference to clean air for smoke opacity (%). It was noticed that boilers operating on furnace oil have larger value of smoke than on natural gas. According to an estimate, at least 3000 different chemicals have been identified in air through sampling of various nature. A term commonly used to describe any harmful chemical or other substance that pollutes

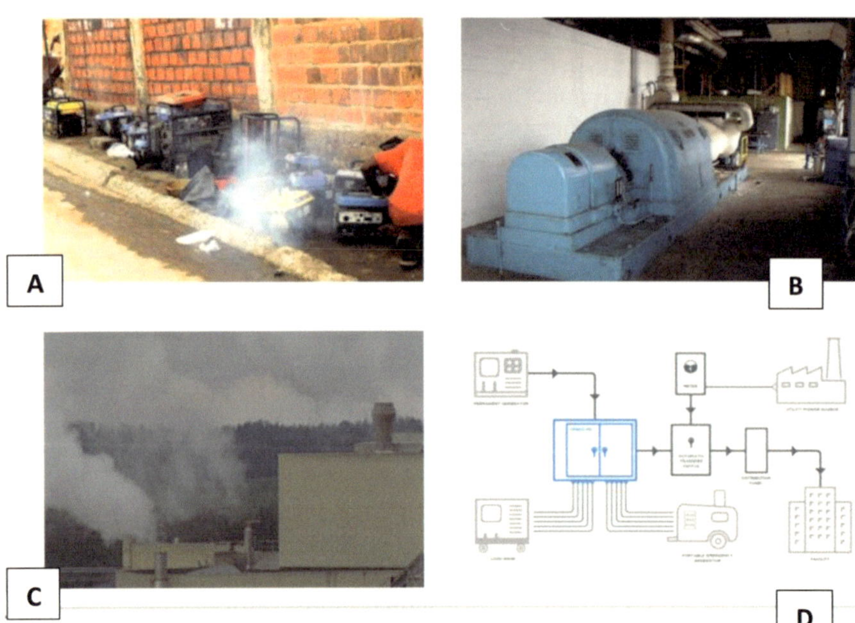

Figure 1.
Presentation of energy sources based upon hydrocarbon fuels: (A) domestic generator; (B) power houses; (C) industrial generators; and (D) energy production.

the air we breathe, thereby reducing its life-sustaining quality is called air pollutant. In principle, air pollutants refer to any chemical substance that exceeds the concentration or characteristics identified as safe for the natural ingredients in the air both by nature or anthropogenically. More strictly, pollutants can be defined a substance which is potentially unsafe to the well-being or health of humans, plant and animal life, or ecosystems. Air pollution is characterized as "the presence of substances in the atmosphere that may adversely affect humans and the environment." It may be a single chemical that is initially produced, or chemicals that are formed by subsequent reactions. According to the World Health Organization (WHO), poor outdoor air caused 4.2 million premature deaths in 2016, of which about 90% were in third world countries. Indoor smoke poses a health threat to 3.0 billion people through heating system and burning biomass, kerosene and coal [10, 11]. Air pollution is linked to a high incidence of respiratory diseases such as cancer, heart disease, stroke and asthma [12]. According to estimates from the American Lung Association, nearly 134 million people are at risk due to air pollution [13]. Although these effects come from long-term exposure, air pollution can also cause acute problems such as sneezing and coughing, eye discomfort, headache, and dizziness [14]. Particles smaller than 10 microns (classified as PM10 or PM2.5 even smaller) pose higher health risks because they can be breathed deeply into the lungs and can enter the bloodstream where air pollutants and nanoparticles have the direct impact on our health [15].

2. Sources of air pollution

Pollutants are commonly classified into solid, liquid, or gaseous substances that are discharged into the air from a fixed or mobile source, then transmit through air, and contribute in chemo physical transformation, and eventually return to the ground. It is impossible to describe the full range of potential sources and actual damage caused by various sources of air pollution but few which are more vulnerable are discussed below:

2.1 Combustion of fossil fuels

Fossil fuels as coal and oil for electricity production and road transportation, add huge amount of air pollutants like carbon dioxide, nitrogen and sulfur dioxide. Sulfur dioxide, oxides of nitrogen and fly ash are produced as main pollutants if coal is used as a fuel. Major pollutants during combustion of oil are oxides of nitrogen and sulfur dioxide, whereas coal emits particulate air pollution to the atmosphere. Similarly, important air pollutants emitted from power station are particulate matter (fly ash and soot) oxides of nitrogen (NO_2 and NO) and sulfur oxides (SO_3 and SO_2) [16, 17]. These pollutants and other closely related chemicals are primarily source for acid rain. When PM is released into the atmosphere due to traffic and industries, these PM scatter the visible part of the sunlight radiation, but the other part of the spectrum particularly inferred and far-infrared, cause the internal heating effect of the air atmosphere below the PM surface. The Sun radiation is heating our air from outside and the traffic and industries from inside. And the PM surface is like a shield or barrier, through the heat diffusion cannot penetrate bidirectional ways.

Volcanic eruption disperses an enormous amount of sulfur dioxide into the atmosphere along with ash and smoke particle sometimes causes the temperature to rise up over the years. Particles in the air, based on their chemical composition, can also have a direct impact of being separated from climate change. They either change the composition or size and may deplete the nutrients biosphere, damage crops, and forests and destroy cultural monuments such as monuments and statues. Many living and non-living sources emit carbon dioxide that contribute largely as pollutant. Carbon dioxide is the most common greenhouse gas, among many others which traps heat into the atmosphere via infrared radiation matching vibrations and causes climate change through global warming. Over the past 150 years, humans have driven enough CO_2 into the atmosphere to make its levels higher than they have been for hundreds of thousands of years. Air pollution in many cases prevents photosynthesis, which has a significant impact on the plants evolution, which has serious consequences for purifying the air we breathe. It also results to form acid rain, atmospheric precipitation in the form of rain, snow or fog, frost, which is released at the time of fossil fuels burning and converted by contact with water vapor in the atmosphere.

2.2 Industrial emissions

Industrial process emits huge amounts of organic compounds carbon monoxide, hydrocarbons, and chemicals into the air. A high quantity of carbon dioxide is the reasons for the greenhouse effect in the air. As the greenhouse gases absorbs infrared radiation from the surface of the planet so its presence is good for the planet. The recent climate change is due to excessive quantity of these gases as well as PM into the atmosphere [18, 19]. Different greenhouse gases contribute differently in global warming due to their unique physical and chemical properties, molecular weight and the lifetime in the atmosphere. A simple working method can calculate the relative contribution of the unit emissions of each gas relative to the cumulative CO_2 unit emissions over a fixed period of time [20, 21]. Therefore, global warming potential (GWP) can be defined as the warming effect of any greenhouse gas relative to CO_2 over a certain period of time. Greenhouse gas emissions from various sources have led to climate change, which has been accompanied by an increase in greenhouse gases [22, 23]. Greenhouse gas emissions change the Climate that is a global issue having significantly negative impacts on economic growth humans, and natural resources [24–26]. The main greenhouse gases (GHGs) and their relative quantities are carbon dioxide, (9–26%), water vapor, H_2O (36–70%), nitrous oxide (3–7%), methane, (4–9%), and other trace gases [27]. Among all the greenhouse gases, CO_2 and CH_4 cause major global surface temperature increase [28]. These gases are emitted by natural and anthropogenically. After carbon dioxide, methane is the second gas that contributes to global warming. Methane has larger impacts as a greenhouse gas than carbon dioxide, with global warming potential (GWP)s 21–25 times higher than CO_2 [29–32].

2.3 Agricultural sources

Agriculture activities often release harmful chemicals like pesticides and fertilizers [33]. Organic matter gradually reduces the water and oxygen in soil during flooding of rice fields; as a result, methane is produce by anaerobic decomposition [34, 35]. Globally methane emission is much lower than CO_2 emissions annually. The concentration of CH_4 in the air is 200 times lesser than carbon dioxide [36] but approximately 20% effects of global warming, because of methane [37, 38]. Naturally it is emitted by marshland [39], termites, wildfires [36], grasslands [36], coal seams [40] and lakes [41]. Human sources of methane include public solid waste landfills coal mine paddy fields oil and gas drilling, pastures rising main sewers, wastewater treatment plants, manure management and agricultural products. Its emission through agriculture sector increased by 11–24% from 2000 to 2010 [33, 42–50].

2.4 Other natural and anthropogenic sources

Natural sources are particulate matter (PM) includes dust produced from the earth's crustal surface, coastal sea salt, form pollens of plant and animal debris [51]. Volcanic eruptions also contribute huge quantities of particles into the environment. Majorly an amount of 3.0 thousand tons of sulfur dioxide emits every day while episodes of great activity. Forest fires of rural areas produce large amounts of all kind of particulate matter including carbon black. Among other sources of natural pollution of air includes lighting in the sky that generate significant quantities of oxides of nitrogen (NO_x); hydrogen sulphide produced from oceans algae and marshy methane. Additionally, concentrations of ozone at ground level, formed because of reaction of nitrogen gases and volatile organic compounds in the presence of sunlight. As far as the human sources are concerned in urban areas, air pollutants come from human-activities, such as cars, trucks, air planes, marine engines, etc. and factories, electric power plants, etc. Nowadays, vehicles on the road constitute the major source of air pollution in the populated areas of countries. Carbon constituted fossil fuels produces carbon monoxide and hydrocarbons whereas NO_x a combination of nitrogen and oxygen gases produced at high temperature. Another very significant thing that road transport accounts a major source of air pollution [52]. It is specified that road transport is the second source of air emissions up to 28.6% after the industrial use of solvents which is 41.4%.

3. Mitigation

Countries, departments and researchers all over the world are dealing for several forms of mitigations for air pollution. In order to restrict global warming, there is a need to take different measures. Important is the addition of more renewable energy sources, substituting gasoline vehicles with zero-emission vehicles as electric vehicles. As an example rapid industrial expansion is China. In china the government is supporting coal-fired power plant. Similarly, in the United States, emission standards setting has improved the air quality, especially in places of worth importance. Contrarily by adding ventilation, using air purifiers, purifying radon gas, running exhaust fans in bathrooms and kitchens and avoiding smoking people can avoid indoor air pollution. While working on a home project, use paint and other products with less volatile compounds. Countries all over the globe have commitments to limit carbon dioxide emissions and other greenhouse gases in the light of Paris Agreement [53, 54] banning hydrophobic hydrocarbons (HFCs) other than chlorofluorocarbon CFCs [55].

3.1 CO_2 sequestering

In this method carbon dioxide is extracted from the air using a solid or liquid adsorbent. Examples of mostly used solid adsorbents include, activated carbon, zeolite, or activated alumina whereas liquid sorbents include, high pH solutions of sodium hydroxide, potassium hydroxide some organic solvents such as monoethanolamine [56, 57]. A method for capturing carbon dioxide from the air includes a number of steps including exposing CO_2 in air to a solution containing an alkali to obtain an alkaline solution that absorbs the carbon dioxide [56].

3.2 Biomass burning

Incomplete combustion of biomass results into production of hazardous gases. The main sources of such emissions are burning of wood, domestic waste, agricultural residues, waste, and charcoal. In developing economies combustion of biomass generally refers to the biofuels combustion for heating, lighting purposes and cooking in small combustion equipment. Because the conditions of burning and types of these fuels vary widely, measures for this category are highly difficult and uncertain to predict.

3.3 Coal mining

Produced of methane by coalification process, and vegetation is transformed into coal by many environmental conditions [58]. The amount of methane gas evolved by mining operations is a function of two main factors: coal depth and coal level [59]. From coal mining, there are four main sources of methane emissions, which are underground coal mines and surface coal mines. These processes account for most of the global emissions of methane from mining. Surface coal mines emit much lower methane as compare to underground coal mines because generally coal mines are at lower rank and capture methane into methane during post-mining operations. Activities of coal mining and processing, continues after operations which emit the methane [60].

3.4 Rice cultivation

Methane emissions through rice production and cultivation can be decreased by selecting proper rice varieties, fertilizers, and water systems. It has been proved that larger total weight rice varieties emit less methane [61, 62]. Fresh straw in the 3 months before transplantation and combined with straw fertilizers before transplantation, plus methane emissions, intermittent irrigation were reduced by 23 and 49%, respectively [63]. Application of potassium fertilizer during flowering period drainage reduce methane emissions.

3.5 Direct utilization of gas

For the production of liquid natural gas and to run leachate evaporators landfill gas can be directly used as fuel. In industrial processes such as kiln operations, boilers, drying operations, and asphalt and cement production landfill methane gas can be used and transported. Natural gas collected from landfills can be transported to local industries directly and use as an alternative or supplementary fuel [64].

3.6 Fuel conversion

Shifting to low-carbon fuels from high-carbon can be comparatively cost-effective principle to reduce the emissions of gaseous because this enhance the efficiency of combustion and reduce the amount of pollutants. In addition, briquette coal and carbon burnout techniques are used in fuel based power plants to minimize the production of pollutants. This pre-combustion method requires almost no hardware changes to the facility and therefore has a lower investment cost. Fuel conversion application to industrial sectors such as the steel, cement and chemical CO_2 emissions can be reduced by 10–20%. There are some essential interrogations about the opportunities that exist for converting fuels in a cost-effective manner. Fuel choices are usually industry dependent, so cost-effective alternatives are limited however, some special opportunities to replace coal-fired boilers with natural gas fired gas-driven steam production; and use natural gas instead of coal to burn blast furnaces [65]. For example, briquette alternative fuels result in a 9–16% increase in fuel costs, although an estimated 26% CO_2 emissions are reduced. Improved fuel efficiency and reduced standard pollutant emissions depends on variable fuel costs which cannot be completely estimated. According to an estimate carbon depletion can save 1.5% of fuel costs. Carbon reductions achieved from ash, by replacing the production of Portland cement is estimated to reduce 144,000 tons of carbon dioxide annually [66].

3.7 Combustion efficiency

Improvement in the current combustion systems have the potential to gear up the energy efficiency. The average thermal performance of current combustion is 32–33% [67]. Control of wasted heat into electricity may result into efficiencies of 45–55%. According to the Department of Energy, the combined energy projects are the main source of greenhouse gases reduction. Technologies like the natural gas combined cycle and combined cycle gas turbine proved for the improving of combustion efficiency and proportionally reduced greenhouse gas emissions and standard pollutant emissions. Additionally, integrated gasification combined cycle system is a step forward to reduce the costs associated with capturing and separating CO_2 from the exhaust stream. Increased operating and fuel costs may be offset by the combined benefits of increased efficiency, reduced pollutants, and credits for emission reductions. An ample evidence that industrial upgradation can reduce greenhouse gas emission, pollutants, and lower operating costs, and current environmental regulations have hindered the adoption of this technology [68]. Air quality regulations determine the operational fuel input rather than power output emission to upgrade of thermal efficiency. However, the environmental agencies provide a guidance document on energy efficiency which begun to address regulatory barriers to improving thermal efficiency [69]. Another source of energy efficiency that can be achieved in the industrial sector is the use of direct fossil fuels. Manufacturing is a major candidate for improving energy efficiency, both of which are achieved through many technological upgrades. Overall, process control and energy management systems for all industries can better control combustion efficiency and fuel use; combined heat and power systems can use waste heat as additional energy; high-efficiency, low-friction motors and drive systems improved the overall efficiency of successfully generating power. In addition to these general categories, various manufacturing industries also have opportunities to improve energy efficiency. Specific industrial sectors with greenhouse gas mitigation potential include cement manufacturing, metal production, refineries, pulp and paper mills, and chemical manufacturing [68].

Combustion efficiency of combustion systems depends on the factors such as type of combustion system, fuel, burner and air fuel ratio for combustion. Significant amount of air pollutants depending on nature of fuel enter into the environment. World health organization (WHO) has provided six listed air pollutants known as classic air pollutants [4]. If coal is used as a fuel, fly ash, sulfur dioxide and oxides of nitrogen are the major pollutant. Combustion of coal produces particulate air pollution whereas in case of oil, sulfur dioxide and oxides of nitrogen are major pollutants emitted to the atmosphere. Similarly, three major air pollutants, particulate matter (fly ash and soot) sulfur oxides (SO_2 and SO_3) and oxides of nitrogen (NO and NO_2) emitted from power station.

Method for calculating efficiency:

$$\text{Efficiency } (\%\, E) = 100 - \Sigma \text{ losses} \quad (1)$$

Losses are as:

1. Temperature flue gas.

2. Moisture in fuel.

3. Combustion of hydrogen.

4. Un-measured losses.

In one of our research work different textile units were examined for stack emissions from boilers and generators. **Table 1** illustrates the results of emissions from boilers. Values of carbon monoxide (CO) were in the range of 0 mg/Nm^3 in CT-Tex to 4903 mg/Nm^3 in HS-Tex. Most of the industries were in compliance of national quality standards of Pakistan, i.e., 800 mg/Nm^3. HS-Tex was exceeding the limit of standards for CO emission. Similarly, **Table 2** represents the gaseous emission of diesel generators. A massive amount of gaseous emissions are produced from generators along with heating which affect the climatic condition at the large scale [4].

Industries	Fuel	CO	CO_2	NO + NO_2	SO_2	H^2
		mg/Nm^3	mg/Nm^3	NO_x g/Nm^3	mg/Nm^3	mg/Nm^3
IP-Tex	Furnace oil	12	221,964	437	3363	0.09
CT-Tex	Natural gas	0	125,321	227	0	0.18
BR-Tex	Natural gas	20	139,857	213	0	1.34
KH-Tex	Natural gas	35	207,428	256	0	1.07
NF-Tex	Natural gas	20	207,625	187	0	1.52
HS-Tex	Natural gas	4903	58,732	121	0	174.2

Table 1.
Gaseous emissions of boilers of textile industries operating with different fuels [70].

Industries	CO	CO_2	NO + NO_2	SO_2	H_2
	mg/Nm^3	mg/Nm^3	mg/Nm^3	mg/Nm^3	mg/Nm^3
IP-Tex	975	21,714	542	80	3
CT-Tex	655	90,000	1342	175	2
BR-Tex	874	131,429	2297	211	2
KH-Tex	572	94,286	2445	265	0.7
NF-Tex	981	86,000	2144	145	3
HS-Tex	1927	76,143	315	65	4

Table 2.
Gaseous emissions diesel generators operation in different industries [70].

4. Conclusion

Quality of life (air) in cities is getting worse as the industrialization, population, energy use and traffic increase. Some air pollutants in larger amount crossing WHO standards, mainly in cities of industrialized countries permitting meaningful statistical trends to air pollutants. The complexity of air pollutants, particularly related to the health impacts in cities, has improved indicators to analyze the accessible monitoring data sufficient for decision making and reporting. Our assessment illustrates that the economic costs of the environmental clash proceeding from sources of combustion in industries tested is potentially excessive. Regarding to the living quality it will be serious concern if no additional control measures were implemented in future. For industrial air pollution there is an immediate need to improve the evaluation and monitoring systems. In cities where strategic planning is not-existing or weak, to improve the quality of air there should be an implementation of environmental management system.

Acknowledgements

Authors are highly thankful to the IntechOpen publishing organization for open invitation to publish a chapter regarding the serious concern of the human atmosphere. Moreover, our especial thanks to Sara Debeuc, who sincerely coordinated to accomplish this chapter. Authors are obliged to Department of Chemistry, University of Gujrat, Pakistan, for providing support of facilities for completing this manuscript.

Conflict of interest

The authors declare no conflict of interest.

Author details

Rabia Munsif, Muhammad Zubair*, Ayesha Aziz and Muhammad Nadeem Zafar
Department of Chemistry, University of Gujrat, Gujrat, Pakistan

*Address all correspondence to: muhammad.zubair@uog.rdu.pk

References

[1] Fanizza C, Baiguera S, Incoronato F, Ferrari C, Inglessis M, Ferdinandi M, et al. Aromatic hydrocarbon levels and PM 2.5 characterization in rome urban area: Preliminary results. Environmental Engineering & Management Journal (EEMJ). 2015;**14**:1583-1593

[2] Klæboe R, Amundsen A, Fyhri A. Annoyance from vehicular air pollution: A comparison of European exposure–response relationships. Atmospheric Environment. 2008;**42**:7689-7694

[3] Kwun SK, Shin YK, Eom K. Estimation of methane emission from rice cultivation in Korea. Journal of Environmental Science and Health. Part A, Toxic/Hazardous Substances & Environmental Engineering. 2003;**38**:2549-2563

[4] Tabaku A et al. Effects of air pollution on children's pulmonary health. Atmospheric Environment. 2011;**45**(40):7540-7545

[5] Weng Z, Mudd GM, Martin T, Boyle CA. Pollutant loads from coal mining in Australia: Discerning trends from the National Pollutant Inventory (NPI). Environmental Science & Policy. 2012;**19**:78-89

[6] Chungsangunsit T, Gheewala SH, Patumsawad S. Emission assessment of rice husk combustion for power production. World Academy of Science, Engineering and Technology. 2009;**53**:1070

[7] Aaheim A, Amundsen H, Dokken T, Wei T. Impacts and adaptation to climate change in European economies. Global Environmental Change. 2012;**22**:959-968

[8] Vaz AIF, Ferreira EC. Air pollution control with semi-infinite programming. Applied Mathematical Modelling. 2009;**33**:1957-1969

[9] Bennett G. 1994. Occupational exposures to mists and vapours from strong organic acids and other industrial chemicals. International Agency for Research on Cancer (IARC). Vol. 54. Geneva, Switzerland: World Health Organization; 1992. p. 336. ISBN: 92-832-1254-1. SWF 65, US $58.50. Elsevier

[10] Karthik S, Sriram A, Vinoth B. Automatic health management system in Urbanized hospitals. Research and Applications: Embedded System. 2018;**1**(2, 3):1-3

[11] Organization, W.H. World Health Statistics 2016: Monitoring Health for the SDGs Sustainable Development Goals. World Health Organization; 2016

[12] To T et al. Progression from asthma to chronic obstructive pulmonary disease. Is air pollution a risk factor? American Journal of Respiratory and Critical Care Medicine. 2016;**194**(4):429-438

[13] Park YM, Kwan M-P. Individual exposure estimates may be erroneous when spatiotemporal variability of air pollution and human mobility are ignored. Health & Place. 2017;**43**:85-94

[14] Lawrence A, Khan T, Azad I. Indoor air quality assessment and its impact on health in context to the household conditions in Lucknow. Global NEST Journal. 2019;**22**:28-41

[15] Idarraga MA et al. Relationships between short-term exposure to an indoor environment and dry eye (DE) symptoms. Journal of Clinical Medicine. 2020;**9**(5):1316

[16] Kim IS, Lee JY, Kim YP. Impact of polycyclic aromatic hydrocarbon (PAH) emissions from North Korea to the air quality in the Seoul Metropolitan

Area, South Korea. Atmospheric Environment. 2013;**70**:159-165

[17] Sivacoumar R, Bhanarkar A, Goyal S, Gadkari S, Aggarwal A. Air pollution modeling for an industrial complex and model performance evaluation. Environmental Pollution. 2001;**111**:471-477

[18] Beauchemin K et al. Use of condensed tannin extract from quebracho trees to reduce methane emissions from cattle. Journal of Animal Science. 2007;**85**(8):1990-1996

[19] Heede R. LNG Supply Chain Greenhouse Gas Emissions for the Cabrillo Deepwater Port: Natural Gas from Australia to California. Climate Mitigation Services. 2006. Available from: http://www.edcnet.org/pdf/Heede_06_LNG_GHG_Anlys.pdf [Accessed: 3 November 2011]

[20] Green HL, Lane WR. Particulate Clouds: Dusts, Smokes and Mists. Their Physics and Physical Chemistry and Industrial and Environmental Aspects. 1957

[21] Abdullah B, Ghani NAA, Vo D-VN. Recent advances in dry reforming of methane over Ni-based catalysts. Journal of Cleaner Production. 2017;**162**:170-185

[22] Absalom H. Meterological aspects of smong. Quarterly Journal of the Royal Meteorological Society. 1954;**80**(344):261-266

[23] Ahrens CD. Meteorology Today. An Introduction to Weather, Climate, and the Environment. 2007

[24] Abbasi T, Abbasi S. Biomass energy and the environmental impacts associated with its production and utilization. Renewable and Sustainable Energy Reviews. 2010;**14**(3):919-937

[25] Bilgen S et al. Global warming and renewable energy sources for sustainable development: A case study in Turkey. Renewable and Sustainable Energy Reviews. 2008;**12**(2):372-396

[26] Yüksel I. Global warming and renewable energy sources for sustainable development in Turkey. Renewable Energy. 2008;**33**(4):802-812

[27] Russell R. The Greenhouse Effect and Greenhouse Gases. Windows to the Universe; University Corporation for Atmospheric Research. 2007

[28] Hansen J et al. Climate change and trace gases. Philosophical Transactions of the Royal Society A: Mathematical, Physical and Engineering Sciences. 2007;**365**(1856):1925-1954

[29] Xiaoli C et al. Characteristics of environmental factors and their effects on CH_4 and CO_2 emissions from a closed landfill: An ecological case study of Shanghai. Waste Management. 2010;**30**(3):446-451

[30] Change I.P.O.C. 2006 IPCC Guidelines for National Greenhouse Gas Inventories. 2006

[31] Todd RW et al. Daily, monthly, seasonal, and annual ammonia emissions from southern high plains cattle feedyards. Journal of Environmental Quality. 2011;**40**(4):1090-1095

[32] Talyan V et al. Quantification of methane emission from municipal solid waste disposal in Delhi. Resources, Conservation and Recycling. 2007;**50**(3):240-259

[33] Dong H et al. Greenhouse gas emissions from swine manure stored at different stack heights. Animal Feed Science and Technology. 2011;**166**:557-561

[34] Huang Y, He Q. Study on the status of output and utilization of landfill gas in China. Journal of Sichuan

University of Science & Engineering. 2008;**1**:117-120

[35] Rodríguez R, Lombardía C. Analysis of methane emissions in a tunnel excavated through carboniferous strata based on underground coal mining experience. Tunnelling and Underground Space Technology. 2010;**25**(4):456-468

[36] Mackie K, Cooper C. Landfill gas emission prediction using Voronoi diagrams and importance sampling. Environmental Modelling & Software. 2009;**24**(10):1223-1232

[37] Lelieveld J, Hoor P, Jöckel P, Pozzer A, Hadjinicolaou P, Cammas JP, et al. Severe ozone air pollution in the Persian Gulf region. Atmospheric Chemistry and Physics. 2009;**9**:1393-1406

[38] Wuebbles DJ, Hayhoe K. Atmospheric methane and global change. Earth-Science Reviews. 2002;**57**(3-4):177-210

[39] Yusuf RO. Methane Emission Inventory and Forecasting in Malaysia. Universiti Teknologi Malaysia; 2013

[40] Cai Y et al. Geological controls on prediction of coalbed methane of No. 3 coal seam in southern Qinshui Basin, North China. International Journal of Coal Geology. 2011;**88**(2-3):101-112

[41] Makhov G, Bazhin N. Methane emission from lakes. Chemosphere. 1999;**38**(6):1453-1459

[42] Zhang G et al. Effect of drainage in the fallow season on reduction of CH_4 production and emission from permanently flooded rice fields. Nutrient Cycling in Agroecosystems. 2011;**89**(1):81-91

[43] Lin H-C, Fukushima Y. Rice cultivation methods and their sustainability aspects: Organic and conventional rice production in industrialized tropical monsoon Asia with a dual cropping system. Sustainability. 2016;**8**(6):529

[44] Karacan CÖ et al. Coal mine methane: A review of capture and utilization practices with benefits to mining safety and to greenhouse gas reduction. International Journal of Coal Geology. 2011;**86**(2-3):121-156

[45] Su S et al. Fugitive coal mine methane emissions at five mining areas in China. Atmospheric Environment. 2011;**45**(13):2220-2232

[46] Wales AD, Allen VM, Davies RH. Chemical treatment of animal feed and water for the control of salmonella. Foodborne Pathogens and Disease. 2010;**7**(1):3-15

[47] Wang S et al. Methane emission by plant communities in an alpine meadow on the Qinghai-Tibetan plateau: A new experimental study of alpine meadows and oat pasture. Biology Letters. 2009;**5**(4):535-538

[48] Shahabadi MB, Yerushalmi L, Haghighat F. Estimation of greenhouse gas generation in wastewater treatment plants–model development and application. Chemosphere. 2010;**78**(9):1085-1092

[49] Guisasola A et al. Development of a model for assessing methane formation in rising main sewers. Water Research. 2009;**43**(11):2874-2884

[50] Etheridge D et al. Historic CH_4 Records from Antarctic and Greenland Ice Cores, Antarctic Firn Data, and Archived Air Samples from Cape Grim, Tasmania. Trends: A Compendium of Data on Global Change. Oak Ridge, Tenn., USA: Carbon Dioxide Information Analysis Center, Oak Ridge National Laboratory, US Department of Energy; 2002

[51] Pénard-Morand C, Annesi-Maesano I. Air pollution: From sources of emissions to health effects. Breathe. 2004;**1**(2):108-119

[52] Festy B. La pollution atmosphérique urbaine: Sources, polluants et évolution. Energies Santé (Paris). 1997;**8**(2):231-241

[53] Fuglestvedt J et al. Implications of possible interpretations of 'greenhouse gas balance' in the Paris agreement. Philosophical Transactions of the Royal Society A: Mathematical, Physical and Engineering Sciences. 2018;**376**(2119):20160445

[54] Michaelowa A et al. Interaction between Art. 6 of the Paris Agreement and the Montreal Protocol/Kigali Amendment. 2019

[55] Petrescu RV et al. NASA sees first in 2018 the direct proof of ozone hole recovery. Journal of Aircraft and Spacecraft Technology. 2018;**2**(1):53-64

[56] Lackner KS et al. Carbon dioxide capture and mitigation of carbon dioxide emissions. Google Patents. 2011

[57] Dietz T, Stern PC, Dan A. How deliberation affects stated willingness to pay for mitigation of carbon dioxide emissions: An experiment. Land Economics. 2009;**85**(2):329-347

[58] Warmuzinski K. Harnessing methane emissions from coal mining. Process Safety and Environmental Protection. 2008;**86**(5):315-320

[59] Gas GAN-CG. Emissions: 1990-2020. Office of Atmospheric Programs Climate Change Division. Washington: US Environmental Protection Agency; 2006

[60] Initiative GGM. Underground Coal Mine Methane Recovery and Use Opportunities. 2008

[61] Xiaohong Z, Jia H, Junxin C. Study on mitigation strategies of methane emission from rice paddies in the implementation of ecological agriculture. Energy Procedia. 2011;**5**:2474-2480

[62] Wassmann R, Hosen Y, Sumfleth K. Reducing Methane Emissions from Irrigated Rice. International Food Policy Research Institute (IFPRI). 2009

[63] Shin Y-K et al. Mitigation options for methane emission from rice fields in Korea. Ambio. 1996:289-291

[64] Yusuf RO et al. Methane emission by sectors: A comprehensive review of emission sources and mitigation methods. Renewable and Sustainable Energy Reviews. 2012;**16**(7):5059-5070

[65] Fernandez CZ, Kulkarni K, Polgar S, Schneider M, Webster SS. A Guide for Small Municipal Utilities

[66] Wilkinson P et al. Public health benefits of strategies to reduce greenhouse-gas emissions: Household energy. The Lancet. 2009;**374**(9705):1917-1929

[67] DOE-ITP. Improving process heating system performance: A sourcebook for industry. In: US Department of Energy, Office of Energy Efficiency and Renewable Energy; 2007

[68] Prindle W et al. Energy efficiency's Next Generation: Innovation at the State Level. Report 2003 (E031). Washington, DC: American Council for an Energy-Efficient Economy; 2003

[69] Prindle J. Videophone and Videoconferencing Apparatus and Method for a Video Game Console. Google Patents. 2003

[70] Zubair M et al. Evaluation of air pollution sources in selected zone of textile industries in Pakistan. Environmental Engineering and Management Journal. 2017;**16**(2)

Chapter 5

Methods to Reduce Mercury and Nitrogen Oxides Emissions from Coal Combustion Processes

Maria Jędrusik, Dariusz Łuszkiewicz and Arkadiusz Świerczok

Abstract

The chapter presents the issue of reducing mercury and nitrogen oxides emissions from the flue gas of coal-fired boilers. The issue is particularly relevant due to the stricter regulations regarding exhaust gas purity. A brief review of the methods for reducing Hg and NO_x emissions has been made, pointing out their pros and cons. Against this background, the results of the authors' own research on the injection of selected oxidants into flue gases to remove both of these pollutants are presented. The injection of sodium chlorite solution into the flue gas (400 MWe lignite fired unit) upstream the wet flue gas desulphurization (WFGD) absorber contributed to the oxidation of both metallic mercury and nitric oxide and enhanced their removal efficiency. The results of tests on lignite and hard coal flue gases indicate that in order to reduce the unfavorable phenomenon of mercury re-emission from WFGD absorbers, in some cases, it is necessary to add selected chemical compounds (e.g., sulfides) to the desulfurization system. The results of field tests for flue gas from lignite (400 MWe unit) and hard coal-fired boilers (195 and 220 MWe units) confirmed the usefulness of oxidizer injection technology to reduce mercury emissions below the level required by BAT conclusions.

Keywords: Hg emissions, NO_x emissions, combustion, industrial pollution, heavy metals

1. Introduction

In nature mercury is present in trace amounts only; due to its toxicity and the ability to join various natural cycles, it poses a threat to human health and life. Mercury exposure, even in small amounts, poses a threat to both people and the environment. A global study commissioned by United Nations Environment Programme (UNEP) confirmed the high environmental impact of mercury, entirely justifying the actions implemented to combat its spread on the international level. In recent years, the European Union has been systematically tightening standards for permissible mercury concentrations in atmospheric air.

According to UNEP data, in 2015 the global emissions from anthropogenic sources amounted to 2220 tons of mercury, accounting for almost 30% of the total atmospheric emissions of mercury. The remaining 70% comes from environmental processes and contemporary natural sources [1]. The technological processes with

IntechOpen

the largest share in mercury emissions are gold production, 38%; coal combustion, 21%; nonferrous metallurgy, 15%; cement plants, 11%; waste incineration plants processing mercury-containing waste, 7%; and combustion of other fuels, including biomass, 3%. Analyzing data on mercury emissions in the respective continents, it can be stated that we find the highest ones in Asia, with about 1084 tons p.a.; in South America, about 409 tons p.a.; Sub-Saharan Africa, 360 tons p.a.; and in the European Union, with 77.2 tons p.a. [1]. Therefore, we can see that the processes of burning fossil fuels form one of the most significant sources of global atmospheric emissions of mercury.

Research on Polish coals [2] demonstrates that the average mercury content in hard coal ranges from 50 to 150 ppb and 120 to 370 ppb in the case of lignite. For comparison, the mercury content of American coals is about 30–670 ppb, with the average content for hard coal of 70 and 118 ppb for lignite. The mercury content in furnace waste indicates that it is mainly found in fly ash and only a small part of it in slag. Literature data indicates that in the result of burning coal, approximately 30–75% of the mercury, contained in the fuel, will be released into the atmosphere [3].

In the process of coal combustion, a number of chemical reactions occur that lead to the decomposition of all chemical compounds containing mercury. In the result of these processes, at a temperature above 600°C, only the metallic mercury Hg^0 in the form of vapor will be present in the exhaust gas [4]. As the exhaust gas is cooled below 540°C [5], this mercury can be oxidized by gas phase components such as NO_2, HCl, SO_2, H_2O, and fly ash, producing various compounds of mercury (**Table 1**).

It was noticed that when burning coals containing significant amounts of chlorine, bromine, or iodine, the concentration of oxidized mercury increases with simultaneous decrease in concentration of metallic mercury. In the process of burning carbons containing chlorine, bromine, or iodine, the process of mercury oxidation is such that during this combustion salts containing chlorine, iodine or bromine is decomposed into HCl, HI, and HBr, whereby 0.5 ÷ 9% of these compounds are further decomposed to CL_2, I_2, and Br_2. These react with metallic mercury to form $HgCl_2$, $HgBr_2$, and HgI_2 salts, respectively, which are stable at high temperatures in vapor form. Oxidized mercury is removed from the flue gas both in dust collectors and in wet and semidry flue gas desulfurization units [6]. However, the efficiency of removal of metallic Hg^0 in the aforementioned devices is low.

No.	Name	Symbol	Boiling point
1.	Mercury	Hg	356.6°C
2.	Mercuric chloride	$HgCl_2$	302.0°C
3.	Mercuric bromide	$HgBr_2$	322.0°C
4.	Mercury(II) iodide	HgI_2	354.0°C
5.	Mercurous oxide	Hg_2O	Decomposes at >100°C
6.	Mercuric oxide	HgO	Decomposes at >500°C
7.	Mercury(I) carbonate	Hg_2CO_3	Decomposes at >130°C
8.	Mercury(II) nitrate	$Hg(NO_3)_2$	Melting point 79°C
9.	Mercury(II) sulfate	$HgSO_4$	Decomposes before reaching liquid phase

Table 1.
Mercury compounds in flue gases from coal combustion processes.

The degree of the removal of mercury and its compounds depends mainly on the degree of transition of metallic mercury to oxidized mercury, with $HgCl_2$ accounting for the main part of oxidized mercury. The value of Hg emissions depends on the combustion process and the method of exhaust gas purification; the mercury removal efficiency in an electrostatic precipitator is 30–40%, while in a wet desulfurization plant, as much as 80–90% of Hg^{2+} (divalent) mercury and mercury adsorbed by the solid phase will be removed, but in the case of elemental Hg^0 mercury, far less is removed, with a removal efficiency of just 26.6% [3].

The proportions between individual forms of mercury in the exhaust gas downstream the boiler depend mainly on the type of furnace and fuel characteristics (mercury, halides, and ash content of coal). The content of halides (fluorine, bromine, iodine, and chlorine) and mercury in fuel has the greatest impact on the amount of Hg^{2+}, while the ash content determines the amount of Hg(p) [7]. For example, the proportions between elemental mercury, oxidized mercury, and ash-bound mercury in flue gas downstream of a pulverized coal boiler are on average 56% (8–94%), 34% (5–82%), and 10% (1–28%), respectively [7]. The type of furnace is not without significance for the mercury speciation in the exhaust gas. Circulating fluidized bed boilers generate the highest amount of Hg(p) (up to 65% of the so-called total mercury Hg^T defined as $Hg^T = Hg^0 + Hg^{2+} + Hg(p)$) due to the extended contact time between gaseous mercury and fly ash and the low temperature of the exhaust gas downstream of the boiler [7].

The European Commission (on July 31, 2017) established conclusions on the best available techniques (BAT) for large combustion plants (LCP). BAT conclusions tighten the regulations related to the emissions from combustion processes, including nitrogen and sulfur oxides, and introduce mercury emission limits (that were not present in the EU till that date). **Table 2** contains the permissible concentrations of mercury and nitrogen oxides in the exhaust gas, resulting from the BAT conclusions. BAT conclusions include ranges of emission limit values for mercury and nitrogen oxides in exhaust gases, with maximum concentration values that will apply from 2021 onwards. Permissible mercury concentrations in exhaust gases resulting from BAT conclusions [8] are referred to as total mercury Hg^T. These values vary depending on the status of the source. For existing sources with a capacity of >300 MW_t, they are 1–4 $\mu g/m^3_{USR}$ for hard coal and 1–7 $\mu g/m^3_{USR}$ for lignite. For new sources with a capacity of >300 MW_t, they are 1–2 $\mu g/m^3_{USR}$ for hard coal and 1–4 $\mu g/m^3_{USR}$ for lignite. Concentrations are converted to standard USR means conditions: (dry gas at a temperature of 273.15 K and a pressure of 101.3 kPa, calculated for oxygen content in the flue gas O_2 = 6 %).

Oxidant	Oxidizing potential, V	Oxidizing potential relative to oxygen
Oxygen, O_2	0.695	1.00
Oxygen radical, O	1.229	1.77
Chlorine, Cl_2	1.360	1.96
Hydrogen peroxide, H_2O_2	1.760	2.53
Ozone, O_3	2.080	2.99
Chlorine (I) anion, ClO^-	0.890	1.28
Chlorate (III) anion, ClO_2^-	0.786	1.13
Hypochlorous acid, HClO	1.630	2.35

Table 2.
Oxidation potentials of oxidants used [31].

BAT conclusions include the range of mercury emission limit values for exhaust gases while specifying maximum concentration values that will apply from August 18, 2021 onwards. The lower values indicate levels that can be obtained using best available techniques, and as long as these values are not required now, it can be expected that existing and new coal units will have to achieve them in near future [8]. This means that users of combustion plants should seek for methods to achieve lower emission levels resulting from the BAT conclusions. The implementation of BAT conclusions thus forms a significant challenge for coal energy in Europe and in particular for the Polish energy sector. The introduction of emission limits also necessitates the addition of Hg^T measurement devices to the pollution monitoring system [8].

BAT conclusions also reduce the permissible levels of nitrogen oxides (NO_x) emissions. For existing sources, fired with hard coal and lignite, with a capacity of >300 MW_t, these amount to 85 (65)–150 mg/m^3, and for new sources with a capacity of >300 MW_t to 50 (65)–85 mg/m^3 in standard conditions.

The above provisions are associated with the need to implement *selective catalytic reduction* (SCR) and *selective non-catalytic reduction* (SNCR) techniques as well as other techniques, including integrated exhaust gas treatment (*multipollutant technologies*), in which a single device is applied to remove at least two pollutants. In this study, we would like to point to the possibility of such integrated flue gas treatment in absorbers of the wet flue gas desulfurization method. The wet limestone method is a common SO_2 removal technology used in power plants both in Europe and worldwide. The desulfurization efficiency of this method ranges from 90 to 95%. This technology is also very popular in Polish conditions, accounting for some 90% of the desulfurization installations.

2. Methods for reducing mercury emissions

2.1 Primary methods

Enrichment of coal prior to the combustion process, e.g., by removing pyrite, can significantly reduce mercury emissions. It is estimated that 65–70% of mercury in Polish coals occurs in combination with pyrite.

Coal enrichment methods are mainly based on physical separation of the mineral substance and involve the use of density differences (gravitational separation) or differences in the wettability of the components (flotation).

One of the methods that do apply dry gravitational separation is the removal of pyrite in purpose-modernized coal mills. The technology is offered by Hansom [9].

Primary methods also include changing the combustion process. For example fluidized bed furnaces to lower the exhaust gas temperature and ash grain composition or using of low emissions burners to lower exhaust gas temperature. Another solution is to replace the coal used for combustion and mixing high Hg and S content coals with those with lower contents of these elements [10]. What is also applied is the addition of halides, in the form of bromine, iodine, and chlorine salts, to the burning coal [11]. The oxidizing properties of these compounds contribute to the increase in the proportion of oxidized mercury in the exhaust gases, which in turn contributes to its more effective retention in existing aftertreatment devices. Unfortunately, these methods cannot guarantee the reduction of mercury to the level required by BAT conclusions.

2.2 Secondary methods

The degree of the removal of mercury and its compounds depends mainly on the degree of transition of metallic mercury to oxidized mercury. Secondary methods consist mainly of removing oxidized mercury adsorbed on ash particles or other adsorbent, e.g., activated carbon, in its form bound with particulates—Hg(p).

An important group of secondary methods are the adsorptive mercury removal methods. They rely on binding of oxidized forms of mercury on the surface of adsorbents. What they use is the affinity of mercury vapors to various adsorbents. The most common adsorber is activated carbon in powdered form (powdered activated carbon). However, due to the limited efficiency of Hg^0 reduction of this typical form of carbon, it is necessary to impregnate this medium with sulfur, iodine, chlorine, or bromine to improve the efficiency of mercury vapor retention. This increases the efficiency of mercury oxidation and its adsorption on PAC particles. Studies demonstrated that ordinary activated carbon can retain up to 80% of mercury in a higher oxidation state but only some 40–50% of elemental mercury. In contrast, carbon impregnated with sulfur, for example, adsorbs over 80% Hg^0 and the iodine impregnated carbon virtually 100% [12].

2.2.1 Injection of activated carbon (PAC) in exhaust gases

Activated carbon is usually injected into the exhaust gas duct before the ESP or fabric filter (**Figure 1**). This technology is used in waste incineration facilities and coal-fired power plants. The effectiveness of this method depends primarily on the type and structure of PAC, the chemical properties of the sorbent surface, the amount of injected coal, and the temperature of the exhaust gas. The main disadvantage of this technology is the increase in the carbon content of ash, which significantly limits the possibilities of ash utilization. Sometimes it can also reduce dust collection efficiency, especially when particles of submicron scale are considered.

To tackle this issue, activated carbon injection downstream the ESP and further exhaust gas purification in the fabric filter are applied (**Figure 2**). However, this makes it necessary to dispose ash from two different locations [13].

Another solution for the injection of activated carbon into exhaust gases is the sorbent injection upstream the air preheater into the zone with a much higher temperature than in the solutions used so far downstream the air preheater or the electrostatic precipitator, i.e., the Alstom Mer-Cure™ technology [14] (**Figure 3**).

2.2.2 The use of systems for catalytic reduction of nitrogen oxides (SCR) for the oxidation of mercury

It was found, based on the research, that in flue gas denitrification installations based on the selective catalytic reduction method, the oxidation of Hg^0 mercury

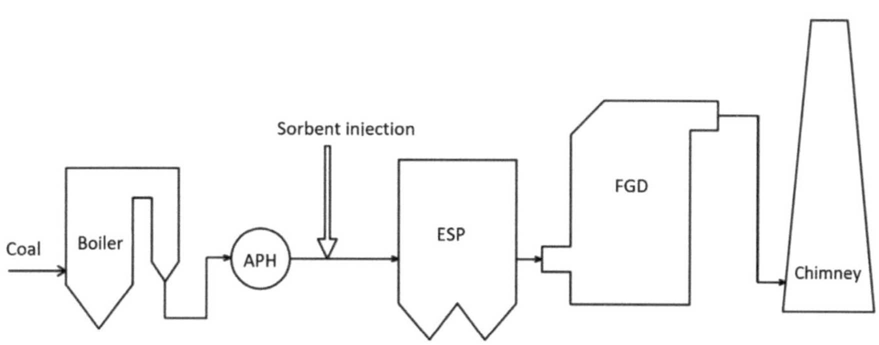

Figure 1.
Diagram of activated carbon injection technology upstream of the ESP; APH—air heater and FGD—flue gas desulfurization installation.

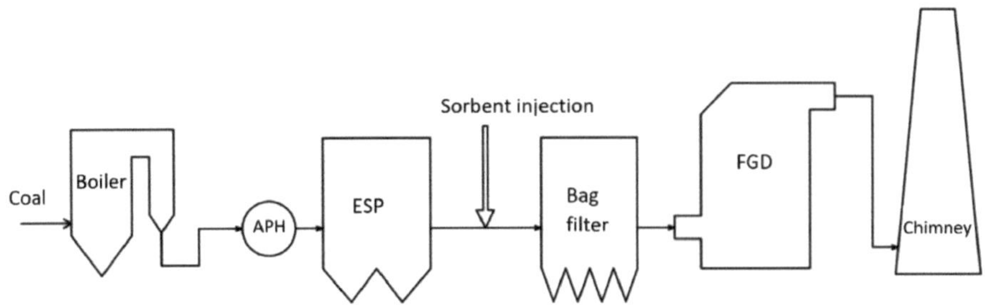

Figure 2.
Diagram of activated carbon injection technology downstream of the ESP; APH—air heater and FGD—flue gas desulfurization installation.

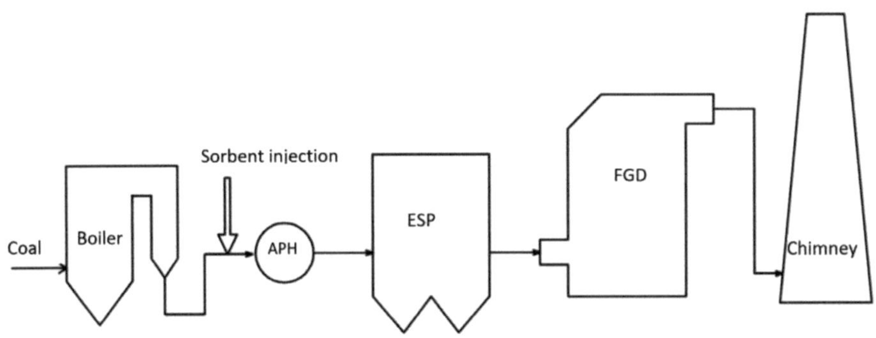

Figure 3.
Diagram of the Mer-Cure™ technology for activated carbon injection; APH—air heater and FGD—flue gas desulphurization installation.

to Hg^{2+} form occurs. The condition for this process, however, is the appropriate chlorine content in the flue gas. Typically, for hard coal, this content proves sufficient to trigger the oxidation process. Important for this process is the fact that the denitrification and oxidation reactions of mercury cannot occur simultaneously, because they depend on the same active centers. Research in industrial conditions indicates that the achievable degree of mercury oxidation is up to 78% [15].

When lignite is burned, the absence of chlorine in the flue gas causes oxidation reactions not to occur. In this case, NH_4Cl injection upstream of the SCR catalyst is proposed to allow mercury oxidation in the catalyst (**Figure 4**). NH_4Cl or NH_4OH injection takes place in a zone with a temperature of about 370–420°C, and then activated carbon is added to the exhaust gas, after which the exhaust gas is directed

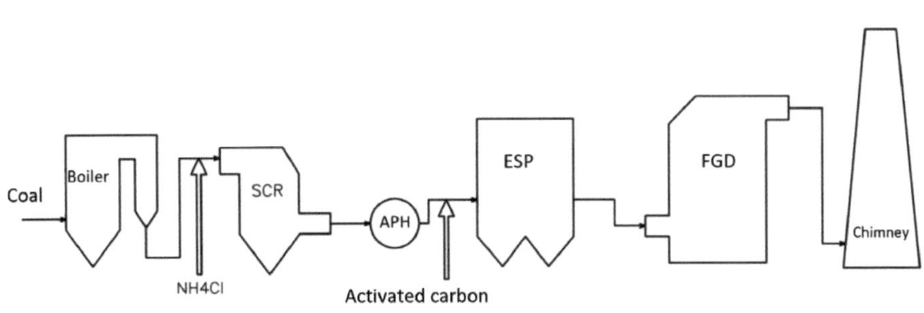

Figure 4.
Diagram of mercury emission reduction technology for lignite-fired boilers: SCR—catalytic flue gas denitrification reactor; APH—air heater; and FGD—flue gas desulphurization installation.

to a dust collector (ESP or fabric filter), and finally to the absorber of the wet desulfurization method [16].

2.2.3 Injection of oxidizing additives and the use of fly ash as the adsorbent

Based on numerous studies [17–23], it was found that with use of chloride additives, it is possible to achieve high efficiency of mercury vapor adsorption on ordinary activated carbon or other sorbents (fly ash) [12, 24].

The proposed method involves the injection of aqueous additive solutions based on chlorite and/or potassium permanganate into the exhaust duct upstream the electrostatic precipitator [25] (**Figure 5**).

The degree of mercury oxidation in this technology depends on numerous parameters; the most important of them are flue gas temperature; flue gas composition, including the SO_2, SO_3, and NO concentrations; pH; and the chemical composition of fly ash. The main oxidized mercury compounds are HgO and Hg $(NO_3)_2$. Part of the oxidized mercury is adsorbed on fly ash particles and as Hg(p) is removed with dust in the ESP unit. The remaining Hg^{2+} mercury in gaseous form is retained in the WFGD absorber and is removed along with the wastewater.

2.3 Removal of oxidized mercury in flue gas purification devices

2.3.1 Removal of mercury in electrostatic precipitators

Tests of mercury content in fly ash upstream of the electrostatic precipitator demonstrate that it is several times higher than the mercury content of coal, which indicates a high sorption capacity of fly ash [26, 27]. The mechanism of mercury adsorption is as follows: in the boiler (temperature of above 1400°C), mercury is in the form of metallic mercury vapors, while the chlorine (HCl) contained in the flue gas activates carbon particles in the ash, and as the flue gas cools down, Hg^0 adsorbs in the chlorinated carbon pores and undergoes oxidation. If there is no HCl (HBr, HI) in the flue gas, there is also no Hg^0 sorption on the ash particles, and the sorption of oxidized $HgCl_2$ mercury is also low.

Research on mercury content in fly ash from hard coal combustion in both pulverized coal and grate boilers indicates a higher Hg content in fine grains. In **Figure 6** we present the results of mercury content testing in individual fractions of fly ash grains from a pulverized coal boiler.

The sorption of mercury and its compounds depends significantly on the flue gas temperature and the content of unburned carbon in fly ash particles. Thus,

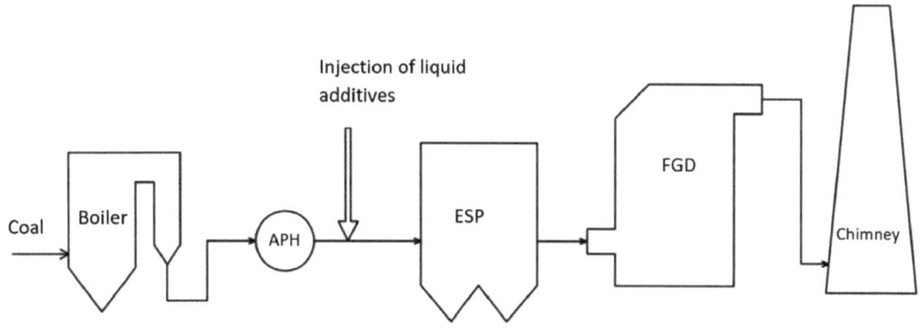

Figure 5.
Diagram of liquid additive injection technology upstream of the ESP: APH—air heater and FGD—flue gas desulphurization installation.

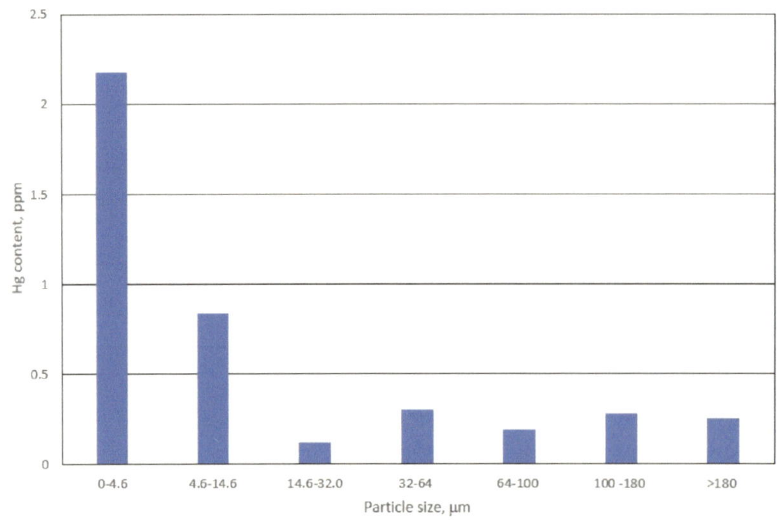

Figure 6.
Mercury content in individual fractions of fly ash from an OP-230 pulverized coal boiler.

the removal efficiency of mercury and its compounds increases with the mercury oxidation efficiency and the increased dust removal efficiency, especially of fine particles.

2.3.2 Removal of mercury in desulphurization installations

2.3.2.1 Mercury removal in absorbers of wet flue gas desulfurization installations

Oxidized mercury compounds contained in the flue gas (mainly the $HgCl_2$) are removed in FGD absorbers, whereas the Hg^{2+} reacts with the sulfides in the exhaust gas, e.g., with H_2S, to form mercury sulfide HgS, which is then precipitated. We also know the phenomenon of mercury re-emission from flue gas desulfurization absorbers. If the sulfide content in the suspension is too low, a chemical reduction of Hg^{2+} to Hg^0 may occur, resulting in higher concentration of metallic mercury downstream the absorber than upstream of it.

It is assumed that the efficiency of removing oxidized mercury in FGD absorbers reaches a value of up to 70%, while it can happen that almost all the oxidized mercury is removed in a dust collector, with only the metallic mercury reaching the absorber [6]. In this case, it is recommended to directly introduce oxidizing additive to the main FGD cycle [28].

2.3.2.2 Removal of mercury in semidry flue gas desulfurization installations

In semidry installations, the desulfurization process of the desulfurization reaction products (waste) remains dry. This process is implemented either by spraying lime milk in the upper part of the reactor (spray dryer) or using the so-called pneumatic reactor, where the sorbent and water are separately fed in its lower part. The resulting dry waste is most often recirculated, and the exhaust gases are dedusted in a fabric filter. The long residence time of sorbent particles in the reactor and the flow of exhaust gas through the filter cake in the bag filter allow for the additional benefit of removing quite a number of impurities, including mercury, provided that an appropriate sorbent is selected.

The semidry method using a pneumatic reactor integrated with a fabric filter for desulfurization of flue gas demonstrated a significant mercury removal efficiency of about 96%, when feeding additional activated carbon together with the primary sorbent (hydrated lime) [29].

3. Methods for reducing nitrogen oxides (NO_x)

Methods for reducing nitrogen oxides from coal combustion in power plants can be divided into two main groups, i.e., the primary and secondary methods. Primary methods rely on the organization of the combustion process in the chamber, primarily through the use of low-emission burners, air staging, exhaust gas recirculation, or reduction of the combustion temperature (fluidized bed boilers). The second group of methods is the secondary method, i.e., the selective catalytic and non-catalytic reduction and oxidative methods.

The latter group of secondary methods is applied in the integrated flue gas cleaning process. The basis for the operation of oxidative methods is the oxidation of sparingly soluble impurities in exhaust gases, i.e., nitric oxide and mercury to soluble forms, and their removal together with SO_2 by means of absorption or condensation [30]. There are many oxidants that are applied in oxidative methods. The most recommended oxidizing agents are ozone (O_3), hydrogen peroxide (H_2O_2), and numerous compounds of chlorine (NaClO, $NaClO_2$, $Ca(ClO)_2$, ClO_2) [31]. Whenever a gaseous oxidant is used, it may be fed directly to the flue gas duct; in the case of liquid oxidants, the conditions necessary for their evaporation should be provided, or, alternatively, they can be used as an additive to the sorption liquid in the absorber [18]. Comparison of the oxidizing potential of individual oxidants with respect to oxygen is presented in **Table 2**.

As you can see, ozone has the highest oxidation potential, and it has the valuable advantage in that it enables oxidation of NO and NO_2 to higher nitrogen oxides, while other oxidants oxidize it predominantly to NO_2 only [31]. The fact that oxidation occurs in the gas phase, which affects the increase in reaction rate, is also significant. Oxidation methods allow for the simultaneous removal of nitrogen oxides, sulfur dioxide, and mercury from flue gases in a single installation, with an efficiency exceeding 90%. Due to the lower operating and investment costs, they form an alternative to the commonly used combination of SCR and FGD. The presence of dust in the flue gas affects the amount of oxidizer used, and therefore a high-performance dust collector should be used upstream of the installation. In the case of commercial pollutant removal installations, ozone is the main oxidizer used for nitrogen oxides. Removal of the reaction products of nitrogen oxides with ozone takes place by means of absorption, for example, by the Lextran [32, 33] and LoTOx methods [34–36]. In Lextran method ozone is added to the flue gas before the absorber feed by mixture of water and catalyst. In LoTOx method, ozone is introduced before FGD absorber.

Another solution is to reduce pollution from flue gas with liquid oxidants. It involves their introduction into flue gas upstream of the wet or semidry flue gas desulphurization installations. Their task is to oxidize both the nitrogen oxide to NO_2 and the metallic mercury to Hg^{2+}. In the case of wet flue gas desulfurization installations, liquid oxidants may also be added to the sorption liquid tank. Hydrogen peroxide [37] is a very popular oxidant used in industry, having the valuable advantage in that it is not as harmful to the environment as chlorine compounds and, at the same time, it is relatively cheap. Exhaust gas treatment with hydrogen peroxide is an extremely promising process. Many researchers around the world are working to improve its effectiveness in relation to the oxidation of nitrogen oxides. Works are carried out on combining the dosing of hydrogen peroxide with metal oxides [38], activating hydrogen peroxide using ultraviolet rays [39], combining H_2O_2 injection with catalysts (Fe-Al, Fe_2O_3, Fe-Ti) promoting the formation of OH^* radicals [40], and using a combination of two oxidants, e.g., $H_2O_2/NaClO_2$ [41]. The results of these experiments are all very promising, and we can expect that future industrial flue gas cleaning installations will apply the presented processes. The achieved efficiency of NO_x and Hg removal from the carrier gas, at least in lab scale tests, is at the level of 90% [42]. Work on the use of sodium chlorite was also carried out on a laboratory and pilot scale [43]. It achieved a removal efficiency of 99% for SO_2 and Hg and 90% for NO_x.

4. Technologies for simultaneous removal of Hg^T and NO_x: authors' own research

As already mentioned, the efficiency of mercury removal in flue gas cleaning installations depends on the speciation of mercury, and the mercury present in the flue gas occurs in both the Hg^0 and the Hg^{2+} forms. Hg^{2+} oxidation increases with the increase in the content of halides (chlorides, bromides, and iodides) in carbon. In the absence of a natural oxidant, as is the case with lignite, liquid oxidative additive can be used for Hg^0 oxidation. Absorbers of the wet flue gas desulfurization plant capture mercury in the Hg^{2+} gas form. In the result of cooperation between the Wrocław University of Technology and Rafako S.A., we developed an Hg emission reduction technology dedicated for hard coal and lignite-fired units. The method involves the injection of sodium chlorite into the exhaust duct upstream the WFGD absorber. In the result of injection of the oxidant, Hg^0 is oxidized to Hg^{2+} and NO to NO_2, and these oxidation products are captured from the flue gas together with SO_2 in the WFGD absorber. The technology has been tested on an industrial scale in a 400 MW_e lignite-fired unit.

4.1 Research on the impact of injection of oxidizer in exhaust gases on the efficiency of Hg and NO_x reduction

The tests were carried out using exhaust gases from a lignite-fired dust boiler (400 MW_e) equipped with a selective non-catalytic NO_x reduction installation, an electrostatic precipitator, and a wet flue gas desulfurization installation. The WFGD absorber is equipped with four levels of sprinkling and a system for feeding adipic acid into the suspension in order to increase the desulfurization efficiency. The test installation for injection of oxidizer (sodium chlorite) was built between the exhaust fan and the fan supporting the WFGD installation. The choice of the additive injection site upstream the booster fan guaranteed very good mixing of the additive with exhaust gases. The mercury content of the fuel during the tests varied between 0.215 and 0.701 mg/kg. A diagram of the installation, along with the location of the measuring points, is shown in **Figure** 7 [44].

As part of the research, we performed continuous measurements of mercury concentration in exhaust gases (using two Gasmet mercury emission monitoring systems) in measuring cross sections located upstream the injection site (A) and in the chimney (C); we carried additional measurements of mercury speciation by the manual method (Ontario-Hydro) at the chimney (C), upstream the WFGD absorber (B), and upstream the oxidative additive injection site (A). Based on the continuous measurements of mercury concentration in the exhaust gas upstream of the absorber and in the chimney, the efficiency of removing mercury from the exhaust gas in the WFGD absorber was calculated with the following formula:

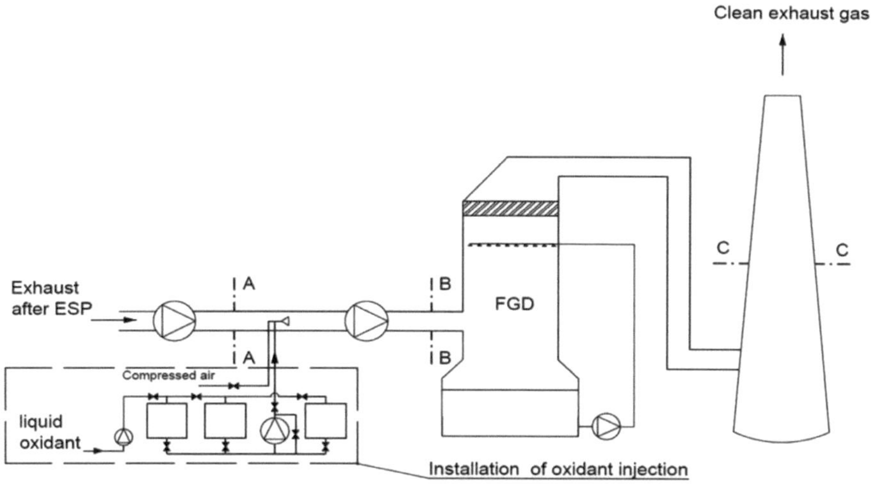

Figure 7.
Diagram of the research installation during tests on lignite flue gas. (A) Measuring cross section before oxidant injection. (B) Measuring cross section downstream the injection site. (C) Measuring cross section in the chimney.

$$\eta_{Hg} = \left(1 - \left({Hg^T}_C/{Hg^T}_A\right)\right) \cdot 100\% \tag{1}$$

where ${Hg^T}_C$ is the mean total mercury concentration in the flue gas in the chimney (C), $\mu g/{m^3}_{USR;}$ and ${Hg^T}_A$ is the mean total mercury concentration in the exhaust gas upstream of the absorber (A), $\mu g/{m^3}_{USR}$.

To determine the NO to NO_2 oxidation degree in a given measurement cross section, the volumetric share of NO_2 in the flue gas in relation to the sum of nitric oxide and nitrogen dioxide (NO_x) was determined. The NO to NO_2 oxidation degree was calculated by means of the relations:

$$\eta^B{}_{NO\rightarrow NO2} = \left({NO_2}^B/{NO_x}^B\right) \cdot 100, \% \tag{2}$$

where ${NO_2}^B$ is the NO_2 concentration in the flue gas in the measurement cross section (B), ppm; and ${NO_x}^B$ is the NO_x concentration in the flue gas in the measurement cross section (B), ppm.

The effectiveness of NO_x removal from the flue gas in the FGD absorber was determined based on the measurement of NO_x concentration (sum of NO and NO_2 calculated as NO_2 [45]) in the cross section located in the chimney (C) and upstream the FGD absorber (A). The NO_x removal efficiency was determined by means of the relation:

$$\eta_{NOx} = \left(1 - \left({NO_x}^C/{NO_x}^A\right)\right) \cdot 100\% \tag{3}$$

where ${NO_x}^A$ is the average NO_x concentration in the flue gas upstream the absorber (A), $mg/{m^3}_{USR}$; and ${NO_x}^C$ is the average NO_x concentration in flue gas in the chimney (C), $mg/{m^3}_{USR}$.

To specify the number of moles of the oxidant to be applied in relation to the moles of nitrogen oxide in the flue gas, a molar ratio X was introduced:

$$X = NaClO_2/NO, mol_{NaClO2}/mol_{NO} \tag{4}$$

Calculation of the molar ratio X was made for the concentration of NO in the flue gas measured in the chimney (C) in the period immediately prior to the oxidant injection.

When the aqueous solution of sodium chlorite is sprayed in the flue gas upstream the absorber, first it evaporates (the temperature of the flue gas during the tests at the oxidant injection site (A) varies from 165 to 170°C) as a result of the reaction of gaseous sodium chlorite (initial pH of sodium chlorite solution was 11.5) with nitric oxide, nitrogen dioxide, and sodium chloride being formed [46]:

$$NaClO_2(l) \rightarrow NaClO_2(g) \tag{5}$$

$$2NO(g) + NaClO_2(g) \rightarrow 2NO_2(g) + NaCl \tag{6}$$

Due to the significant share of moisture in the flue gas (from 28 to 29%), there were very good conditions for the formation of nitric and nitrous acids [47]:

$$2NO(g) + NaClO_2(g) \rightarrow 2NO_2(g) + NaCl \tag{7}$$

The nitric acid formed in the flue gas reacted with the metallic mercury and oxidized it to the form Hg^{2+} (mercury(II) nitrate), which increases Hg^T removal efficiency from flue gas [43, 46]:

$$Hg^0(g) + 4HNO_3(g) \rightarrow Hg(NO_3)_2 + 2NO_2(g) + 2H_2O(g) \tag{8}$$

Because flue gas contains acidic gases such as SO_2, HCl, and HF, they can be absorbed by oxidant droplet and drop its pH before evaporation which caused the release of ClO_2 [48]. Chlorine dioxide can directly oxidized NO and Hg^0; additionally emission of chlorine radical is possible, which enhanced Hg^0 oxidation [15, 19]:

$$5ClO_2^- + 4H^+ \rightarrow 4ClO_2(g) + 2H_2O\ (l) + Cl^- \tag{9}$$

$$2ClO_2(g) + NO(g) \rightarrow NO_2(g) + Cl^* \tag{10}$$

$$Cl^* + Hg^0 \rightarrow HgCl \tag{11}$$

$$Hg^0\ (g) + ClO_2\ (g) \rightarrow HgCl\ (g) + O_2\ (g) \tag{12}$$

In such a complicated gas mixture as flue gases from lignite combustion, the presented mechanism can occur simultaneously. For example, the efficiency of NO to NO_2 oxidation and the removal of Hg^T and SO_2 during the tests carried out in a lignite-fired power plant (sodium chlorite fed to the exhaust gas prior to the FGD absorber) are shown in **Figure 8**.

The efficiency of Hg^T removal and oxidation of nitrogen oxides in exhaust gases depend on the stream of injected sodium chlorite to exhaust gases, which is illustrated by the molar ratio X. Changes in total mercury concentration in exhaust gases in the chimney (C) and NO, NO_2, and NO_x downstream the sodium chlorite injection site (B) are illustrated in **Figure 9**. The undoubted advantage of the presented method is the almost immediate reaction of the entire system to the injected sodium chlorite. An increase in the amount of injected additive (series I < series II) causes an immediate decrease in the Hg^T concentration in the chimney and an increase in the NO_2 concentration in the exhaust gas downstream the injection site. The

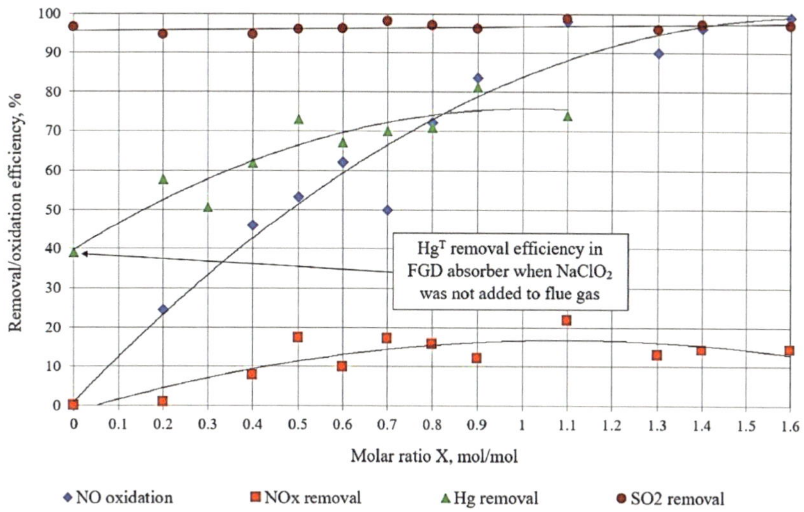

Figure 8.
Oxidation NO to NO_2, NO_x, SO_2, and Hg^T removal efficiency in function of molar ratio X.

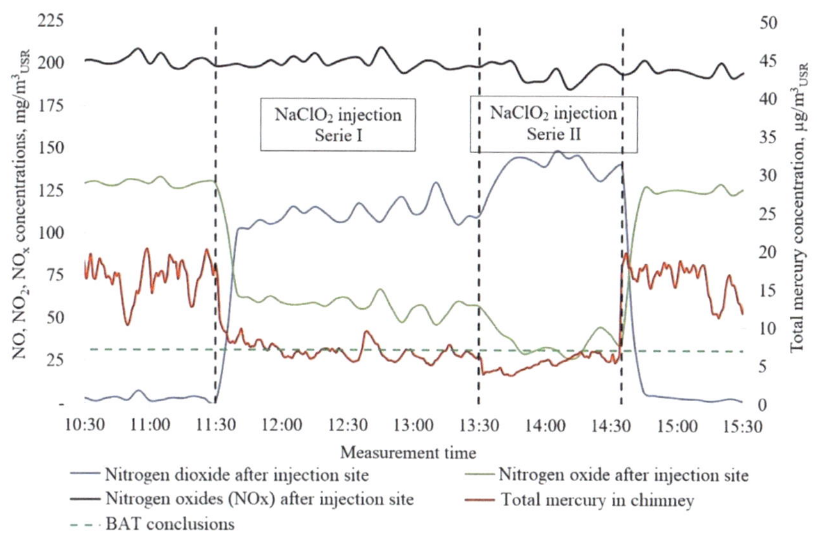

Figure 9.
NO, NO_2, and NO_x concentrations in the flue gas downstream the injection site (B) and Hg^T concentration in the chimney (C).

Hg^T concentration in the chimney during the presented tests was below the level required by the BAT conclusions, i.e., <7 μg/m^3_{USR}.

Sodium chlorite injection into flue gas upstream of the WFGD absorber caused an increase in Hg^{2+} concentration in the flue gas, which translated into the efficiency of mercury removal. Unfortunately, in some cases, the increase in Hg^{2+} concentration in the exhaust gas intensified the phenomenon of re-emission [44].

4.2 Increased Hg removal efficiency by limiting re-emissions

The phenomenon of re-emission consists in chemical reduction of the Hg^{2+} absorbed in the suspension to the elemental Hg^0 mercury emitted back into the atmosphere [49]. Sulfite ions (SO_3^{2-}), acting as a reducing agent, are responsible for this phenomenon [50]:

$$Hg^{2+} + SO_3^{2-} + 3H_2O \rightarrow Hg^0 + SO_4^{2-} + 2H_3O^+ \quad (13)$$

$$Hg^{2+} + HCOOH + 4H_2O \rightarrow Hg^0 + HCO_3 + 3H_3O^+ \quad (14)$$

In FGD installations, where the addition of organic acids (formic, adipic and other) serves increasing the SO_2 removal efficiency, the following reaction takes place (14) [50]. Dosing organic acids increases the concentration of Ca^{2+}, which improves the efficiency of SO_2 removal from the exhaust gases. Many researchers also reported the clear effect of sulfite concentration in the suspension on Hg^0 re-emission. Generally, an increase in SO_3^{2-} concentration increases the re-emission [51–53].

The re-emission phenomenon is assessed on the basis of measurements of mercury concentration in exhaust gas both upstream and downstream the WFGD absorber. In order to find out the nature of the re-emission phenomenon, research was carried out on a lignite-fired unit. We assumed that the concentration of total mercury in the cross section (C) was higher than in the cross section (B) the phenomenon of mercury re-emission from the FGD absorber was present, and the intensity of this phenomenon was described using re-emission rate:

$$\eta_{re-emission} = \left(\left(Hg^T{}_C - Hg^T{}_B\right)/Hg^T{}_B\right) \cdot 100\% \quad (15)$$

An example of variations in total mercury concentration in exhaust gases in the period when re-emission occurred is presented in **Figure 10**.

The observed phenomenon of mercury re-emission from the absorber lasted for approx. 4 h. Based on the analysis of the presented graphs, we calculated the degree of mercury re-emission according to Eq. (5); the calculation results are presented in **Figure 11**.

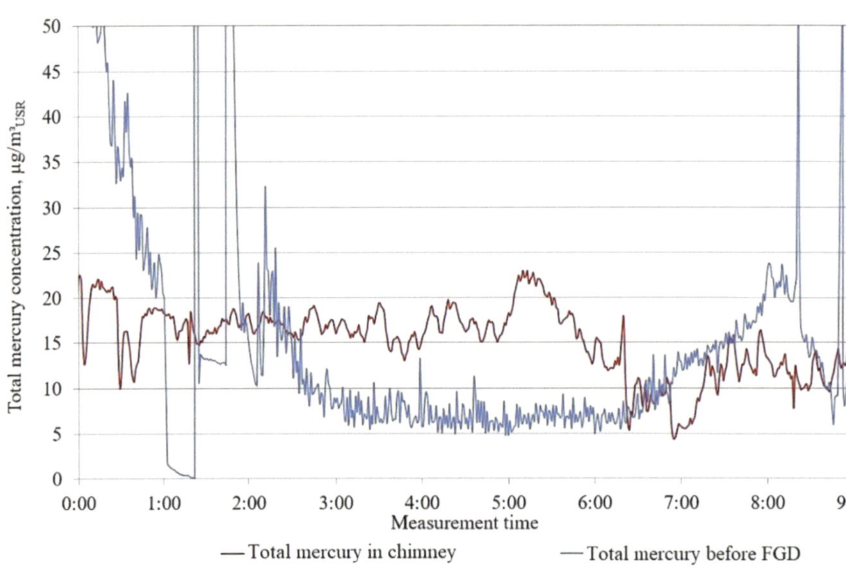

Figure 10.
Total mercury concentrations in flue gas upstream the WFGD absorber (B) and in the chimney (C).

The observed degree of re-emission from the WFGD absorber reached 220%. In order to explain the mechanisms of this phenomenon, the results of the re-emission degree were compared with the operating parameters of the unit and the WFGD (**Figure 12**). Mercury re-emission occurred when the absorber operating parameters changed, and the pH and ORP proved to be the most significant of them. A detailed description of the parameters affecting the intensity of the phenomenon of re-emission from the WFGD absorber is presented in the publication [44].

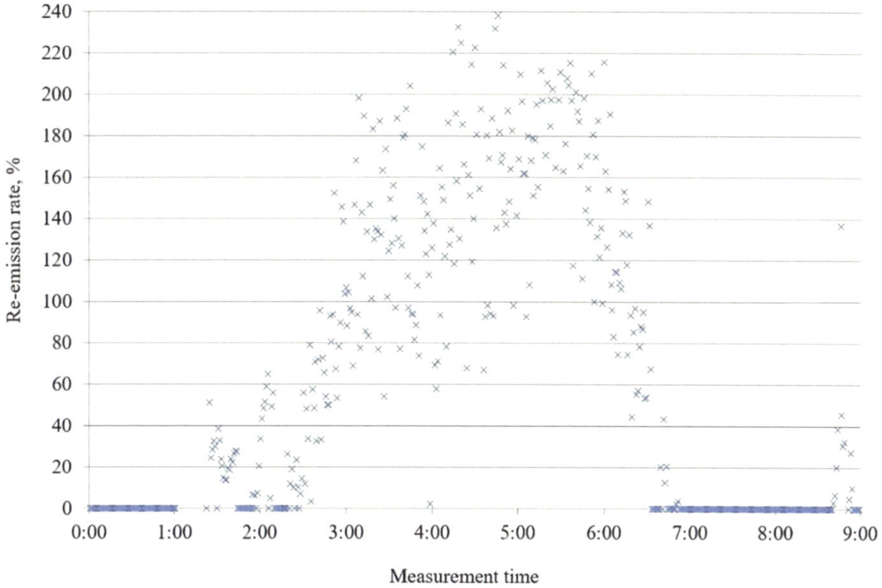

Figure 11.
The degree of mercury re-emission from the WFGD absorber during measurements for a lignite-fired unit.

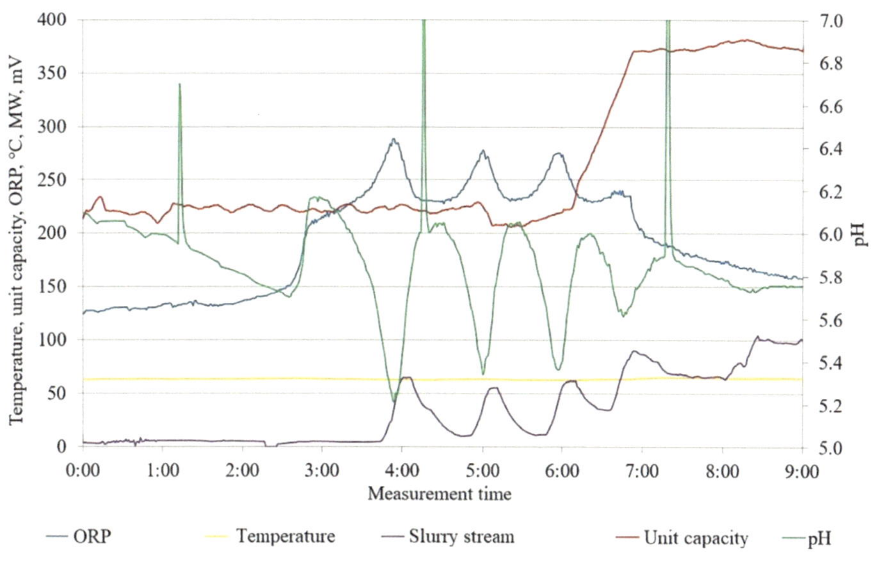

Figure 12.
Parameters of unit and WFGD absorber operation during measurements for a lignite-fired unit.

Research demonstrated that re-emission can be reduced by changing the absorber's operating parameters. We noticed that an increase in suspension temperature and pH increased re-emission, while the increase in chloride concentration in the suspension and the intensity of air flow through the suspension reduced it [54]. At the same time, numerous studies indicate that significant reductions of Hg^0 re-emission can be obtained by adding various additives [53–55]. The most common are simple additions of NaHS and Na_2S organic sulfides with a more complex structure. The goal is always the same, i.e., to remove from the solution (suspension) Hg^{2+} by formation HgS, which prevents re-emission. The effect of adding sodium sulfide (Na_2S) to the suspension circulation in the WFGD absorber was studied for a lignite-fired unit, and the results are presented in **Figure 13**. 4 m^3 of 10% solution of sodium sulfide were pumped directly into the tank under the absorber. In this way, the mercury concentration in the exhaust gas was reduced below the level required by the BAT conclusions (7 $\mu g/m^3_{USR}$) for a period of approx. 4 h.

The phenomenon of mercury re-emission from the WFGD absorber is not always identifiable on the basis of measurements of total mercury concentration in exhaust gases. Hard coal tests were carried out for the WFGD absorber, purifying flue gas from two units with a capacity of 195 and 220 MW_e. During the tests, both boilers operated at maximum power. Prior to the tests, measurements were performed with the Ontario-Hydro method revealing that the absorber is experiencing metallic mercury re-emission. The results of these measurements are presented in **Figure 14**.

The total mercury removal efficiency in the flue gas treatment installation (electrostatic precipitator and WFGD) was 72.4%. Mercury bound with the ash was virtually completely removed in the ESP. The flue gas downstream of the boiler contained a small amount of metallic mercury only (1.73 $\mu g/m^3_{USR}$), which was a result of the high concentrations of halides in the fuel (Cl (0.110 ÷ 0.211%), Br (0.008 ÷ 0.011%), F (0.002 ÷ 0.004%)). The concentration of metallic mercury in the exhaust gas upstream of the absorber was lower than downstream the absorber, which meant that the absorber was the source of mercury re-emission. The total mercury removal efficiency in the ESP was 56.2% and another 36.9% in the WFGD absorber. Due to

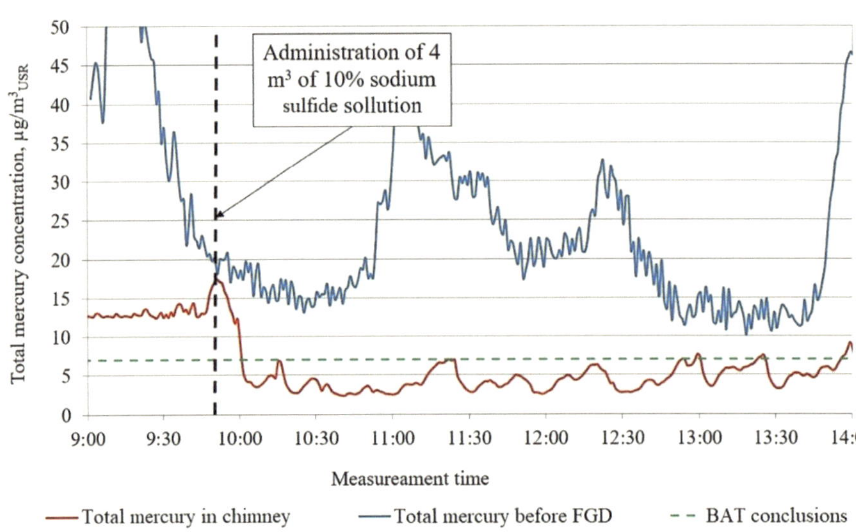

Figure 13.
Total mercury concentration in the chimney and upstream the WFGD absorber after a one-time injection of $4m^3$ of sodium sulfide (10%).

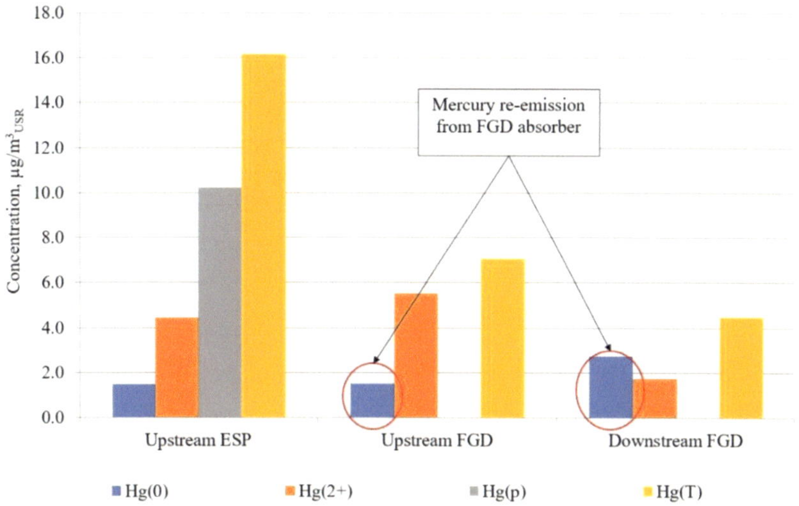

Figure 14.
Comparison of mercury concentration in flue gas for hard coal tests.

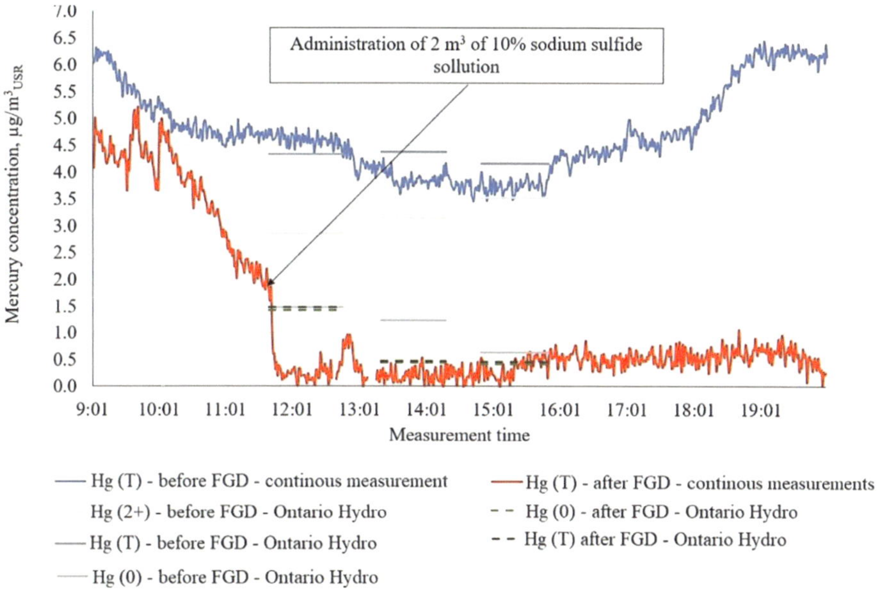

Figure 15.
Measurement results of mercury concentration in flue gas upstream and downstream the WFGD absorber (continuous and Ontario-Hydro measurements) during the addition of Na_2S.

the fact that the proportion of oxidized mercury upstream the WFGD absorber is significant, sodium sulfide was fed to the absorber to reduce mercury emissions in the flue gas in the chimney. In **Figure 15**, we present the results of measurements of mercury concentration in exhaust gas upstream and downstream the WFGD absorber, during dosing of sodium sulfide. Measurements were carried out with two continuous emission monitoring systems and the Ontario-Hydro method.

The total mercury concentration in the exhaust gas before the administration of sodium sulfide was 4.3 $\mu g/m^3_{USR}$, and after the addition of sodium sulfide, the concentration of total mercury in the exhaust gas dropped to 0.45 $\mu g/m^3_{USR}$. The mercury removal efficiency for the exhaust gas in the WFGD absorber amounted to 25.5% without the addition of sulfide and increased to 90.5% after applying the additive. To sum up, due to the content of halides in coal, a considerable amount of Hg^{2+} is present in hard coal exhaust gas, which can be effectively removed in WFGD, as long as the phenomenon of re-emission is controlled.

5. Summary

The chapter presents selected issues related to Hg and NO_x emissions from coal combustion processes, in the aspect of regulations related to limiting permissible emissions of pollutants, as contained in the BAT conclusions. The review of methods applied to reduce mercury emissions demonstrates that the specific technology should be selected individually for each facility considered. There is no single, universal, cost-effective solution. In order to choose an effective method for reducing mercury emissions, it is first and foremost necessary to hold the knowledge of the speciation of mercury in the exhaust gas downstream the boiler. In the case of low concentration of oxidized mercury, there are no devices that can be installed in order to secure sufficient limiting of mercury emissions. In such a case, one should first consider the solutions that consist in supplementing the exhaust gas with additives to oxidize the metallic mercury first.

Among the methods used for denitrification of exhaust gases, attention has been given to oxidative methods, which form an opportunity to simultaneously reduce NOx and Hg emissions. The results of the author's own research in industrial conditions confirmed the usefulness of injection of the oxidant (sodium chlorite) to the exhaust gas upstream the WFGD absorber to reduce mercury emission. Under favorable conditions for lignite flue gases, up to 70% Hg removal efficiency was achieved, coupled with 17% NO_x removal efficiency and an unchanged SO_2 removal efficiency. Whenever there is the phenomenon of re-emission of mercury from the WFGD absorber, appropriate measures must be undertaken to limit it. Again, test results on lignite and hard coal exhaust gas indicate that it is possible to reduce re-emissions to such an extent, as to ensure compliance with emission standards in line with BAT conclusions.

By using mercury oxidation technologies with simultaneous application of flue gas purification devices ($DeNO_x$, $DeSO_x$, and dedusting) and effectively combating re-emissions, we can achieve total mercury concentrations at the level required by BAT conclusions, i.e., in the order of 1–7 (4) $\mu g/m^3_{USR}$.

List of abbreviations

APH	air (pre)heater
BAT	best available techniques
ESP	electrostatic precipitator
FGD	flue gas desulphurization
PAC	powdered activated carbon
SCR	selective catalytic reduction
WFGD	wet flue gas desulphurization

Author details

Maria Jędrusik*, Dariusz Łuszkiewicz and Arkadiusz Świerczok
Faculty of Mechanical and Power Engineering, Wrocław University of Science and Technology, Wrocław, Poland

*Address all correspondence to: maria.jedrusik@pwr.edu.pl

References

[1] Global Mercury Assessment. UN Environment Programme. Geneva, Switzerland: Chemicals and Health Branch; 2018. ISBN: 978-92-807-3744-8

[2] Wojnar K, Wisz J. Rtęć w polskiej energetyce. Energetyka. 2006;**4**:280-283 (in Polish)

[3] Pavlish JH et al. Status review of mercury control options for coal-fired power plants. Fuel Processing Technology. 2003;**82**:89-165

[4] Niksa S, Fujiwara N. The impact of wet flue gas desulfurization scrubbing on mercury emissions from coal-fired power stations. Air & Waste Management Association. 2005;**55**:970-977

[5] Gale T, Lani B, Offen G. Mechanisms governing the fate of mercury in coal-fired power systems. Fuel Processing Technology. 2008;**89**:139-151

[6] Nguyen YV, Pessione GF. A three-year assessment of mercury mass balance from Lambton's coal fired boilers equipped with FGD and SCR. In: Power Plant Air Pollution Control Symposium, Baltimore, August 28-31, 2008

[7] Zhang L, Wang S, Wu Q, Wang F, Lin C, Zhang L, et al. Mercury transformation and speciation in flue gases from anthropogenic emission sources: A critical review. Atmospheric Chemistry and Physics. 2016;**16**:2417-2433

[8] EU Parliament Decision. Commission Implementing Decision (EU) 2017/1442 of 31 July 2017 Establishing Best Available Techniques (BAT) Conclusions, under Directive 2010/75/EU of the European Parliament and of the Council, for large combustion plants (notified under document C(2017) 5225)7

[9] Available from: http://www.hepaus.com/images/PDFs/hep_FPCS_MAPS_Desander_h.pdf

[10] Guidance on Best Available Techniques and Best Environmental Practices Coal-Fired Power Plants and Coal-Fired Industrial Boilers [Internet]. 2016. Available from: https://pdfs.semanticscholar.org

[11] Vosteen B et al. Hg-Oxidation durch Chlor, Brom und Iod in Braunkohle-Kesseln. In: 51. Kraftwerkstechnisches Kolloquium—Annual Conference of the Energy Industry and Power Plant Industry, 22-23 October 2019. Dresden: International Congress Center; 2019

[12] Oleniacz R. Oczyszczanie gazów odlotowych ze spalania odpadów niebezpiecznych. Inżynieria Środowiska. 2000;**5**(2):85-94 (in Polish)

[13] Lindau L, Durham M, Bustard J, Cameron M. Mercury: Myths and realities. Modern Power Systems. 2003;**3**:30-32

[14] Available from: www.alstom.com/mercury-emissions-technology

[15] Krotla K. Wykorzystanie systemów katalitycznego oczyszczania spalin do redukcji emisji rtęci—Podstawy teoretyczne i przykłady z praktyki. In: VI Forum dyskusyjne ENERGOPOMIAR, Tatrzańska Łomnica, 16-19 kwietnia. 2013 (in Polish)

[16] Nakamoto T, Kato Y, Nagai Y, Neidig K. SCR Catalyst, A Low Cost Technology for Mercury Mitigation, Hitachi Paper_Neidig_100213, Library: Mitsubishi Hitachi, Technical Papers

[17] Jak W. EPA, No. ICR: Information Collection Request for Electric Utility Steam Generating Unit Hg Emissions, Information Collection Effort. 1858; 1999

[18] Krzyżyńska R, Hutson ND. Effect of solution pH on SO_2, NO_x, and Hg removal from simulated coal combustion flue gas in an

oxidant-enhanced wet scrubber. Journal of the Air & Waste Management Association. 2012;**62**:212-220

[19] Wilcox J et al. Mercury adsorption and oxidation in coal combustion and gasification processes. International Journal of Coal Geology. 2012;**90-91**:4-20

[20] Hower JC et al. Mercury capture by native fly ash carbons in coal-fired power plants. Progress in Energy and Combustion Science. 2010;**36**:510-529

[21] Wilcox J. A kinetic investigation of high-temperature mercury oxidation by chlorine. The Journal of Physical Chemistry. 2009;**113**(24):6633-6639

[22] Senior CL, Johnson SA. Impact of carbon-in-ash on mercury removal across particulate control devices in coal-fired power plants. Energy & Fuels. 2005;**19**:859-863

[23] Cauch B, Silcox GD, Lighty JAS, JOL W, Fry A, Senior CL. Confounding effects of aqueous-phase impinger chemistry on apparent oxidation of mercury in flue gases. Environmental Science & Technology. 2008;**42**(7):2594-2599

[24] Gostomczyk MA, Jędrusik M. Doświadczalna instalacja do redukcji emisji SO_2, NO_x i rtęci ze spalin kotłowych. Archiwum Energetyki. 2008;**38**(2):97-104. (in Polish)

[25] Jędrusik M, Świerczok A, Krzyżyńska R. Usuwanie rtęci w elektrofiltrach. Przemysl Chemiczny. 2014;**93**(11):1885-1888 (in Polish)

[26] Lee SJ et al. Speciation and mass distribution of mercury in a bituminous coal-fired power plant. Atmospheric Environment. 2006;**40**:2215-2224

[27] Zhang L et al. Mercury emission from six coal-fired power plants in China. Fuel Processing Technology. 2008;**89**:1033-1040

[28] US 2002/0068030A1. Method for Controlling Elemental Mercury Emission. Patent US 2002/0068030A1, June 6, 2002

[29] Knura P. Półsucha metoda odsiarczania spalin z zastosowaniem reaktora pneumatycznego zintegrowanego z filtrem tkaninowym (metoda RP + FT)—Kierunki rozwoju technologii, potencjał i możliwości. In: VI Forum dyskusyjne ENERGOPOMIAR, Tatrzańska Łomnica, 16-19 kwietnia 2013 (in Polish)

[30] Carpenter AM. Advances in Multi-Pollutant Control. IEA Clean Coal Centre; 2013. Available from: https://www.usea.org/sites/default/files/112013_Advances%20in%20multi-pollutant%20control_ccc227.pdf

[31] Ozonek J. Analiza procesów wytwarzania ozonu dla potrzeb ochrony środowiska. Lublin: Państwowa Akademia Nauk; 2003 (in Polish)

[32] Udasin S. Firm to test out technology for purifying emissions. The Jerusalem Post. 2012. Available from: www.jpost.com/Sci-Tech/Article.aspx?id=269347

[33] Lextran. Lextran Retrofit/Upgrade Solution: Effectively Controlling the Emissions and the Expenses. Israel: Lextran; 2012. Available from: www.lextran.co.il/objects/Retrofit-upgrade20-9-12.pdf

[34] Omar K. Evaluation of BOC's Lotox process for the oxidation of elemental mercury in flue gas from a coal-fired boiler. United States; 2008. DOI: 10.2172/993830

[35] Jarvis JB, Day AT, Suchak NJ. LoTOx™ process flexibility and multi-pollutant control capability. In: Combined Power Plant Air Pollutant Control Mega Symposium, Washington, DC, USA, 19-22 May 2003. Pittsburgh, PA, USA, Air and Waste Management Association, Paper 147. 2003

[36] A pioneering NO_x removal technology for the power industry. Modern Power Systems. 2015;**35**(5):32-33

[37] Crapsey K. Eco Power Solutions Multi-Pollutant Emissions Control Systems. Northfield, IL, USA: The McIlvaine Company; 2012. Available from: www.mcilvainecompany.com/Universal_Power/Subscriber/PowerDescriptionLinks/Kevin%20Craspey%20-%20Eco%20Power%20Solutions%208-17-12.pdf

[38] Liu YX, Wang Q, Yin YS, Pan JF, Zhang J. Advanced oxidation removal of NO and SO_2 from flue gas by using ultraviolet/H_2O_2/NaOH process. Chemical Engineering Research and Design. 2014;**92**:1907-1914

[39] Huang XM, Ding J, Zhong Q. Catalytic decomposition of H_2O_2 over Fe-based catalysts for simultaneous removal of NO_x and SO_2. Applied Surface Science. 2015;**326**:66-72

[40] Ding J, Zhong Q, Zhang S. Catalytic efficiency of iron oxides in decomposition of H_2O_2 for simultaneous NO_x and SO_2 removal: Effect of calcination temperature. Journal of Molecular Catalysis A: Chemical. 2014;**393**:222-231

[41] Zhao Y, Hao RL, Guo Q, Feng YN. Simultaneous removal of SO_2 and NO by a vaporized enhanced-Fenton reagent. Fuel Processing Technology. 2015;**137**:8-15

[42] Hao R, Zhao Y, Yuan B, Zhou S, Yang S. Establishment of a novel advanced oxidation process for economical and effective removal of SO_2 and NO. Journal of Hazardous Materials. 2016;**318**:224-232

[43] Krzyżynska R, Hutson ND. The importance of the location of sodium chlorite application in a multi pollutant flue gas cleaning system. Journal of the Air and Waste Management Association. 2012;**62**(6):707-716

[44] Jędrusik M, Gostomczyk MA, Świerczok A, Łuszkiewicz D, Kobylańska M, et al. Fuel. 2019;**238**: 507-531. DOI: 10.1016/j.fuel.2018.10.131

[45] Polish Standard PN93/Z-04009/06. Air Purity Protection. Examination of the Content of Nitrogen and Its Compounds. 1993 (in Polish)

[46] Lee HK, Deshwal BR, Yoo KS. Simultaneous removal of SO_2 and NO by sodium chlorite solution in wetted-wall column. Korean Journal of Chemical Engineering. 2005;**22**:208-213. DOI: 10.1007/BF02701486

[47] Sun Y, Hong X, Zhu T, Guo X, Xie D. The chemical behaviors of nitrogen dioxide absorption in sulfite solution. Applied Sciences. 2017;7(4):377. DOI: 10.3390/app7040377

[48] Hao R, Wang X, Liang Y, Lu Y, Cai Y, Mao X, et al. Reactivity of $NaClO_2$ and HA-Na in air pollutants removal: Active species identification and cooperative effect revelation. Chemical Engineering Journal. 2017;**330**:1279-1288. DOI: 10.1016/j.cej.2017.08.085

[49] Ochoa-Gonzales R et al. Control of Hg^0 re-emission from gypsum slurries by means of additives in typical wet scrubber conditions. Fuel. 2013;**105**:112-118

[50] Heidel B, Hilber M, Scheffknecht G. Impact of additives for enhanced sulfur dioxide removal on re-emissions of mercury in wet flue gas desulfurization. Applied Energy. 2014;**114**:485-491

[51] Keiser B, et al. Improving Capture of Mercury Efficiency of WFGDs by Reducing Mercury Re-Emission. US8110163B2. 2012

[52] Wo J et al. Hg^{2+} reduction and re-emission from simulated wet flue gas desulfurization liquors. Journal of Hazardous Materials. 2009;**165**(2-3):1106-1110

[53] Omine N et al. Study of elemental mercury re-emission in simulated wet scrubber. Fuel. 2012;**91**:93-101

[54] Wang Y, Liu Y, et al. Experimental study on the absorption behaviors of gas phase bivalent mercury in Ca-based wet flue gas desulfurization slurry system. Journal of Hazardous Materials. 2010;**183**:902-907

[55] Tang T, Xu J, Lu R, Wo J, Xu X. Enhanced Hg^{2+} removal and Hg^0 re-emission control from wet flue gas desulfurization liquors with additives. Fuel. 2010;**89**(12):3613-3617

Section 3

Composition and Measurement of Emissions

Chapter 6

Carbon Soot Polymer Nanocomposites (CSPNCs): Production, Surface Morphological, Glass Transition Temperature Phenomenon and Optical Properties

Rakhi Tailor, Yogesh Kumar Vijay and Minal Bafna

Abstract

The present chapter covers the production and properties of carbon soot nanoparticles (CSNPCs) and their doped carbon soot polymer nanocomposites (CSPNCs). The first part of this chapter will provide a brief introduction of carbon soot, its morphology, production and synthesis methods. The second part will explain the investigation of carbon soot nanoparticles by flame deposition method and their properties. The third part will provide a short knowledge on polymer nanocomposites (PNCs) and their processing methods. The last part will illustrate the production of carbon soot polymer nanocomposites by solution casting method and their important properties. At the end, the chapter concludes with future scopes.

Keywords: carbon soot (CS), carbon soot polymer nanocomposites (CSPNCs), almond soot (AS), mustard soot (MS), energy band gap (E_g)

1. Introduction

Carbon particulate matters are small spherical particles in the range between 100 and 500 Å of diameter and synthesized by the combustion process [1, 2]. We will study these spherical particles by means of electron microscopy. X-ray diffraction technique (XRD) reveals that each soot particle contains thousands of crystallites and carbon sheet atoms with the graphite structure [3]. Researchers observed that the soot particles are aggregates of a number of tiny spherical particles, named primary particles [4, 5]. The chemical study of soot particles reveals that these particles are composed of elements other than just carbon. Carbon soot particles are composed of 1% hydrogen [1, 2] and up to 3% in weight of polycyclic organic matter [6]. In 1960, Lindsay collected soot samples by the combustion of 16 different hydrocarbons and extracted them with cyclohexane [7]. Carbon soot has several kinds of sources such as burning of oils, biomass fuels or any deliberate fires. The investigation of soot in a flame is a complete process, composed of many mechanisms, and it depends on the composition of fuels [8]. In 1984, Melton suggested that carbon soot is an insoluble solid substance and roughly consists of eight carbon atoms and a single hydrogen atom [9]. Joyce found that soot is hard amorphous carbon particles in microscopic size [10]. Researchers found that soot particles are small in size but spherical hard particles in the range of 10–35 nm. These nanoparticles basically consist of carbon atoms arising from the fuel combustion at high temperature with a condition when the oxygen amount is insufficient to convert all the fuel in to CO_2 and H_2O [11, 12]. Soot particles can aggregate in soft and slippery secondary particles by their structure of agglomeration having average size of 120 nm or greater than 400 nm [13, 14].

2. Carbon soot

2.1 Carbon soot morphology

The structure of soot is difficult to define to a properly satisfactory extent. Soot obtain from combustion process and pyrolysis of hydrocarbons are composed of primary spheres agglomeration with a range of 10–15 nm in a diameter for a broad type of operation conditions [15].

The distribution of size mainly depends upon the primary spheres. Some researchers recorded that mustard soot has parallel physical and chemical characteristics to form the different types of soot, and it has been found that the primary particles could be as small as in a diameter [16].

Morphology of carbon soot particles from different field flames have been carried out using scanning electron microscopy (SEM) and transmission electron microscopy (TEM). The primary particles are composed of agglomerates like chain of individual spheres [17, 18]. The internal structure of soot particles is analyzed by high-resolution phase contrast electron microscopy [19].

According to the chemical study of soot, carbon is a main component in soot, but hydrogen and minor constituents can be present. The carbon hydrogen mole ratio of premature soot particles (primary soot particles) obtained by hydrocarbon fuels is equal to one or less than one, but this ratio for mature soot particles (aggregate soot particles) is high or approximately equal to 10 as compared to premature (C/H) ratio [18]. Agglomerated particles produced by the combustion method have been found to be in the range of 100 nm to 2 μm [20, 21]. The nanosize of soot particles is dependent on various conditions such as the production of sampling, operations and type of fuels. The term nanostructure of the soot is used to characterize the graphene layer plane dimensions its relative orientation and tortuosity [22, 23].

2.2 Production and destruction method of soot

Soot particles basically consist of carbon, oxygen and hydrogen, and some other constituents such as nitrogen, zinc, sulfur, etc. are also observed [24, 25]. In **Table 1**, the typical composition of soot obtained from the lubricant oil and soot contained are shown [26]. Soot in individual particles or spherical agglomerates is composed of many primary particles with approximately 10–50 nm in diameter [27]. The size of soot particles increases due to agglomeration because some particle diameters are up to 200 nm, and in this process, the soot-free radical character decreases.

The formation of soot is a complex process because this involves several physical and chemical steps and is still regarded as only partially understood. Soot particles

Soot composition	Soot content (%)
Carbon	90
Oxygen	4
Volatile content	6

Table 1.
Soot percentage composition.

obtained from different sources such as various fuels and combustion are expected to be approximately similar in composition and size, suggesting the same processes of formation and growth constraints. Many authors have presented a model of soot formation from liquid or vapor phase combustion, which mainly comprises or is classified in to five distinct and normal identified processes, namely pyrolysis, nucleation, surface growth, coalescence and agglomeration as presented in **Figure 1** [18, 28, 29].

Carbon soot production is also visualized as being a fuel thermal stability function [30]. Some authors also investigated that in the pyrolysis all hydrocarbons yield acetylene, which is the most precursor of species by incomplete combustion [31–33]. This is contributed to its low thermal stability where it easily breaks in to carbon solid and hydrogen. A well-defined and accepted theory for the formation of soot is explained by Haynes [18].

2.3 Investigation of carbon soot nanoparticles

Carbon nanosoots have been produced by two different edible oils by flame deposition techniques in two different environments.

2.3.1 Formation of soots in natural environment (without water tub)

At room temperature, the flame deposition method is a well known, cost-effective and easiest method to produce carbon nanosoot from different oils (**Figure 2**). In this technique, we take two diyas or "clay lamp": one filled with almond oil and the other one is filled with mustard oil. Dip a cotton wick in both diyas and put it on the floor at some distance (approx. 40 cm). Light up cotton wick. To collect carbon soot particles, put another diya in reverse at some height with the help of other layers of diyas and make the arrangement.

In this arrangement, the distance between the tip of the wick and collector diyas' surface is approximately 5 cm. The distance has been optimized by carrying out the pyrolysis process at various source collector distances. The flame stabilizes after 1–2 min and then the almond oil soot and mustard oil soot are allowed to deposit on the collector. For 2 h, the almond and mustard oils were burned, and then the generated soots were scrapped off using a spatula on butter paper and then samples were packed in air-tight bottles with proper labeling.

2.3.2 Formation of soots in water tub environment (with water tub)

In this section, the procedure of soot production from almond and mustard oil at a room temperature is the same as above mentioned (see Section 2.4.1). The only change made in this section is to introduce a water tub. The entire setup is arranged in a water tub to provide the sample surroundings extra moisture (**Figure 3**). After completion of the processes, all soot samples were stored in air-tight bottles and labeled (**Figure 4**).

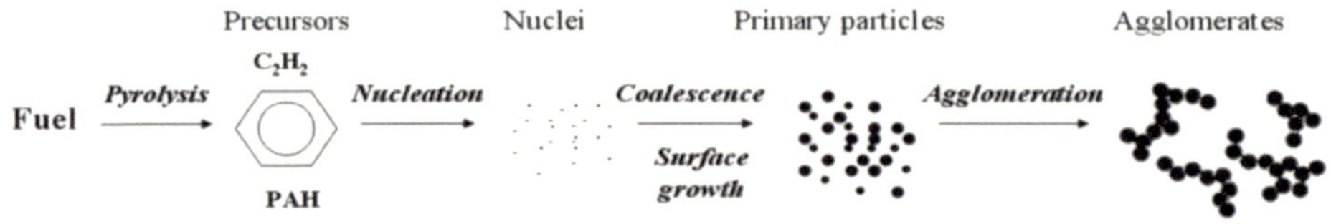

Figure 1.
Soot formation steps from fuel.

Figure 2.
Carbon soot-producing setup in natural environment.

2.4 Properties of carbon soot nanoparticles (CSNPCs)

2.4.1 X-ray diffraction analysis of almond soot (AS) and mustard soot (MS) nanoparticles

2.4.1.1 XRD analysis of almond oil soots

The X-ray diffraction spectra in **Figure 5** are from almond soot without and with water tub environment. Samples show carbonaceous soot. The two theta values at 24.74 and 42.13° are identified for natural environment (without water tub) as graphite peaks. These peaks correspond to plane (002) and (101), respectively. Similarly, the sample with water tub got two prominent peaks. A high-intense peak at 24.04° and a low-intense peak at 43.69° appeared. As shown in the figure, the peaks of samples in the presence of water tub are with greater intensities as compared to samples in natural environment. Due to this property, its crystallinity is not lost. A high-intense peak at 24.74 and 24.04^0 for natural and with water tub soot indicates the presence of large amount of amorphous carbon. Two peaks are at 42.13 and 43.69° and indicate the presence of low-quality carbon nanomaterial. All these peaks denote the presence of multiwalled graphite carbon nanotubes.

2.4.1.2 XRD analysis of mustard oil soots

The XRD profiles of carbon soot that are obtained without and with water tub (**Figure 6**), respectively, are quite similar in their pattern, both showing two

Figure 3.
Carbon soot-producing setup in water tub environment.

Figure 4.
Picture of generated soot samples for characterization.

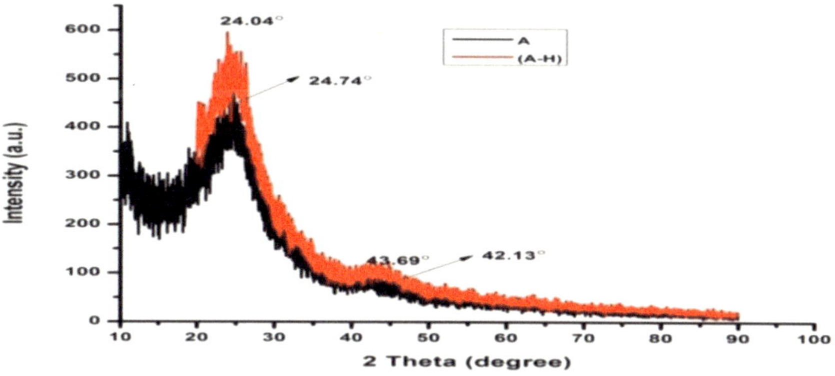

Figure 5.
XRD spectra of almond soot without (A) and with water tub (A-H) environment.

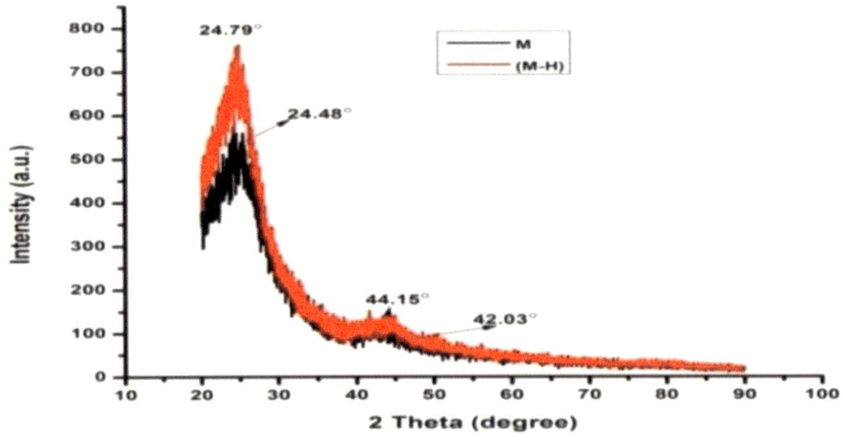

Figure 6.
XRD patterns of mustard soot without (M) and with water tub (M-H) environment.

prominent broad diffused peaks. The sample with water tub exhibits higher intensity suggesting that the surrounding of water or creation of humid environment makes it less amorphous. For the soot samples without water tub, high-intense Bragg diffraction peak is found at $2\theta = 24.78°$ and a low-intense peak at 44.15° for (002) and (100) reflections, respectively. These reflections suggest that the soots obtained by this method have a good extend to layer formation. Similarly, in the case of soot sample with water tub, a high-intense peak at 24.48° and low-intense peak at 42.03 show that crystallinity is not lost due to oxidative acid treatment, and these two peaks at 24.48 and 24.70° are a high-intensity broad peak, which indicates the presence of large amount of amorphous material in the soots. The low-intensity peaks at 42.03 and 44.15° are an indication of the low quality of carbon nanoparticles present in soots with and without water tub, respectively, and these observations are in quite agreement with the result reported by other researchers [34, 35].

2.4.2 Field emission scanning electron microscopy analysis of almond soot (AS) and mustard soot (MS) nanoparticles

2.4.2.1 Field emission scanning electron microscopy analysis of almond oil soots

The almond soot material deposited on glass surface was investigated by scanning electron microscopy for without and with water tub conditions. The particles are very small occurring nonindividually and individually. Synthesized soot particles in flame deposition method break up to form other small substances. For the first condition, surface morphology is seen to be nonuniform (**Figure 7a**), and there are several grains that look like carbon nanotubes, but for the second condition, it is seen to be uniform and particle size is small comparatively (**Figure 7b**). The SEM images for both conditions show carbon soot particles of average size approximately 50 nm. The formed soot particles show morphology of agglomerated clusters for without water tub sample and uniformly distributed carbon nanoparticles for with water tub sample.

2.4.2.2 FESEM analysis of mustard oil soots

Mustard oil is burned in two different environment to obtain soot by a flame deposition process in which oil breaks up to form other substances. The SEM images of obtained particles describe the synthesized carbonaceous particles from burning of oil and are a mixture of elemental carbon and oxygen, a variety of hydrocarbons. The surface morphology of the deposited carbon obtained is seen to be nondistinguishable in natural environment sample (**Figure 8a**) and distinguishable in water tub environment sample (**Figure 8b**).

The prepared soot particles are agglomerated in a range between 50 and 100 nm, but most particles are about 50 nm in size. As shown in all SEM images, agglomerated soot particles with well-defined boundaries are arranged in the form of a chain (carbon necklace).

2.4.3 Chemical composition analysis of almond soot (AS) and mustard soot (MS) nanoparticles

2.4.3.1 Energy dispersive X-ray study of almond oil soots

Energy dispersive X-ray of almond oil soot particles is presented for natural and with water tub condition. The spectra show the presence of carbon and oxygen. For the first condition (**Figure 9a**), the EDAX analysis indicates the soot to consist 92.10% weight of carbon and 7.9% weight of oxygen. For the second condition (**Figure 9b**), it consists of 87.86% carbon and 12.14% weight of oxygen. **Table 2** shows the percentage composition of elements carbon and oxygen obtained from

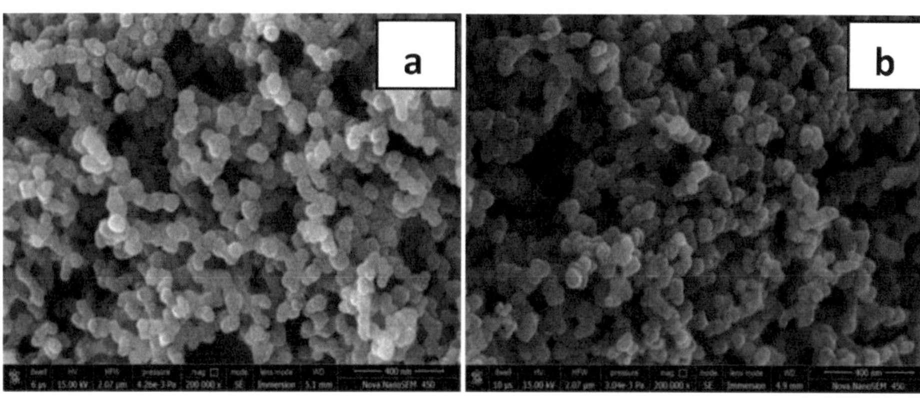

Figure 7.
Field emission scanning electron micrograph of almond oil soot (a) natural environment and (b) with presence of water tub.

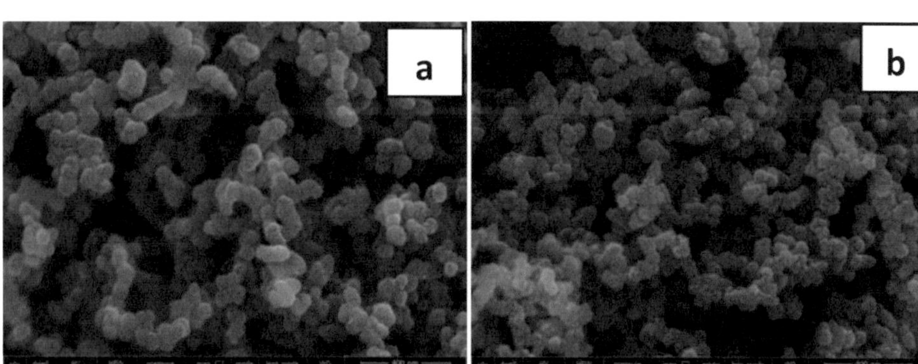

Figure 8.
Field emission scanning electron micrograph of mustard oil soot (a) natural environment and (b) with presence of water tub.

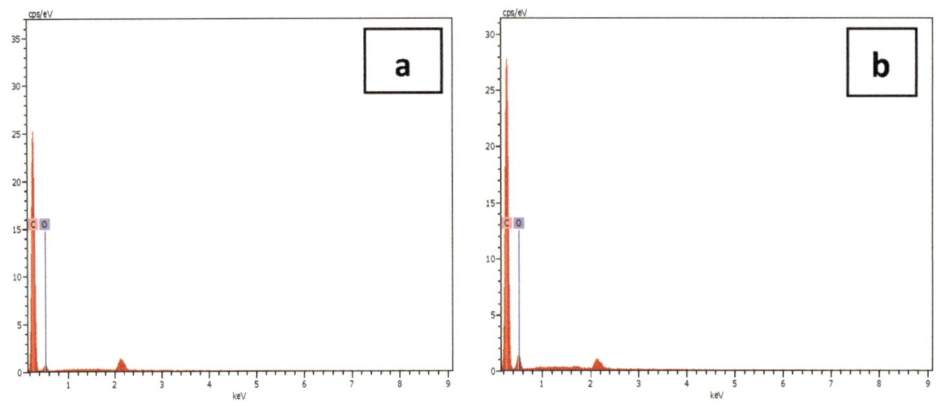

Figure 9.
EDAX spectra of almond oil soot in (a) without and (b) with water tub environment.

Samples of soot particles	Percentage of carbon and oxygen elements	
	Carbon element (C%)	**Oxygen element (O%)**
Almond oil soot without water tub	92.10	7.90
Almond oil soot with water tub	87.86	12.14
Mustard oil soot without water tub	83.37	16.63
Mustard oil soot with water tub	90.81	9.19

Table 2.
Elemental composition of almond and mustard oil soot in natural and with water tub environment.

EDAX spectra of different samples in different conditions. The spectra reveal that almost 85% of the sample contains pure carbon and remaining 15% oxygen, confirming the absence of any other external impurities. This AS sample is electrically nonconducting, and for EDAX analysis, this sample is coated with gold to convert it into electrically conducting. Due to of this, an extra small peak is observed for gold at approximately 2 keV in EDAX spectra.

2.4.3.2 EDAX study of mustard oil soots

The elemental analysis of synthesized mustard soot was performed using energy dispersive X-ray (EDAX). The spectra show the presence of carbon and oxygen for both conditions. The composition of soot aggregates from the EDAX analysis shows the soot to consist of about 83.37% weight of carbon and 16.63% weight of oxygen element for first condition (**Figure 10a**), and similarly for second condition (**Figure 10b**), it consists of about 90.81% weight of carbon and 9.19% weight of oxygen as shown in **Table 2**. The result shows the product of the flame deposition of mustard oil to be composed of mainly almost (80–90)% carbon and remaining (10–20)% oxygen.

The EDAX of soot samples indicates the presence of no other except carbon and oxygen. This MS sample is electrically nonconducting, and for EDAX analysis, this sample is coated with gold to convert it into electrically conducting. Due to of this, an extra small peak is observed for gold at approximately 2 keV in EDAX spectra.

2.4.4 Fourier transform infrared spectroscopy analysis of almond soot (AS) and mustard soot (MS) nanoparticles

2.4.4.1 Fourier transfer infrared analysis of almond oil soots

FTIR transmission spectra for without and with water tub samples are shown in **Figure 11**.

A peak at 3435 cm^{-1} is for water O—H stretch, peaks at 2918 and 2919 cm^{-1} are for C—H stretch or carboxylic acid O—H, and peaks at 2850 cm^{-1} are for C—H stretch or (—C—H) aldehyde. A weak peak for both soot spectra at 1617 cm^{-1} is detected for C=C aromatic stretch. Two bands are at 1384 and 1377 cm^{-1} stretch for $-CH_3$ group. Some detected bands at low wave numbers 1048, 668 cm^{-1} and 1047, 705 cm^{-1} indicate the presence of SP^2 and SP^3 aromatic clusters of amorphous carbon soot.

2.4.4.2 FTIR analysis of mustard oil soots

FTIR spectroscopy analysis was carried out to identify the chemical structure of carbon soot nanoparticles as well as the presence of any functional group, and the obtained spectra are shown in **Figure 12,** respectively.

FTIR spectra of Mustard Oil soot without and with water tub environment conditions, shows that in between 1500 to 4000 cm^{-1} range spectra resemble in their profile and full spectra remain in the range 400 to 4000 cm^{-1}.

The broad vibration band at 3435 cm^{-1} is attributed to the OH stretching of the carboxylic acid group. It also indicates the presence of absorbed moisture of OH group in the samples, and it is therefore more intense for with water tub sample as compared to without water tub as expected. Another vibration band at 2919 and 2852 cm^{-1} is due to symmetric and antisymmetric C—H bond stretching. The band at 1725 cm^{-1} is assigned to C=O stretching vibration associated with the presence of carboxylic acid group. A band centered at 1619 cm^{-1} with a shoulder at 1384 cm^{-1} may be due to (C=C) stretching vibrations. A band at 1217 cm^{-1} is associated with CO stretching vibrations. A band centered at 1047 cm^{-1} corresponds to C—N stretch. Some obtained bands at low wave numbers 826, 738 and 597 cm^{-1} indicate the presence of small (SP^2 and SP^3) clusters.

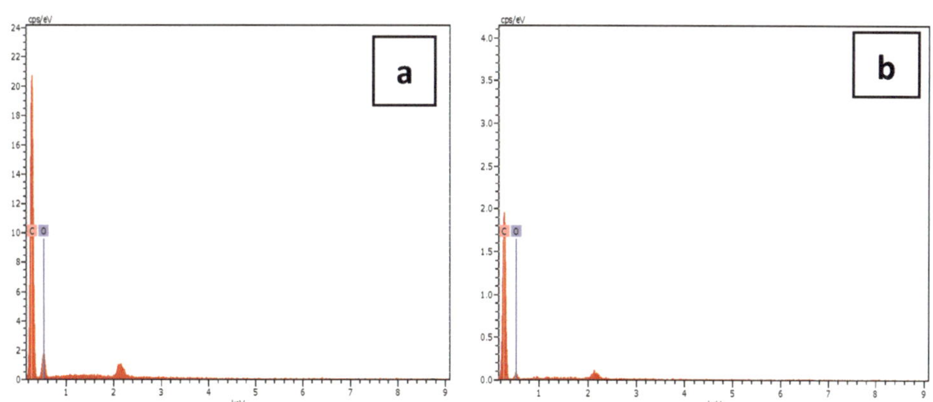

Figure 10.
EDAX spectra of mustard oil soot in (a) without and (b) with water tub environment.

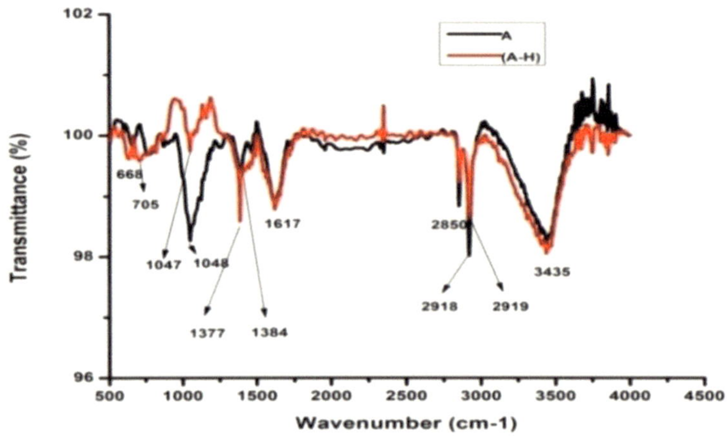

Figure 11.
FTIR spectra of almond soot in without (A) and with water tub (A-H) environment.

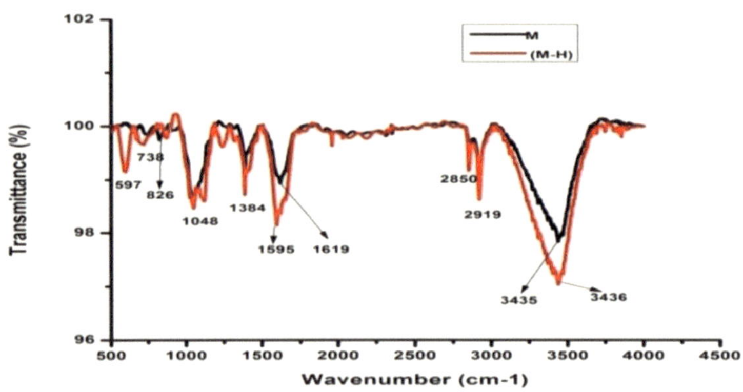

Figure 12.
FTIR spectra of mustard oil soot in without (M) and with water tub (M-H) environment.

These observations are in good agreement with the earlier reported ones [35–37]. So the IR studies directly prove the presence of carboxylic acid groups and clusters in synthesized soot.

3. Polymer nanocomposites (PNCs)

The nanomaterials can be divided into two types: one is nanostructured materials and another one is nanophase materials. Normally, bulk materials are made by grains (agglomerates) with the nanometer range. Nanocomposites refer to materials existing of minimum two phases with one dispersed in another and making a 3D network. The dispersion medium is called matrix and the dispersed phase is called filler. If the host matrix of nanocomposite is polymer, then the resulting nanocomposite is called polymer nanocomposite (PNC). PNC belongs to the nanostructural class material. If filler has higher dimensions, the resulting composite refers to polymer composite (PC). PNCs are different to other PC materials; their filler consists at the nanoscale [38, 39]. According via, PNC's particles exists on nanoscale measure and particles with nanoscale arrangements has a polymer host matrix with nanoparticles. PNCs have been shown to possess the classic thermal and mechanical properties [40, 41]. Nanoparticles are in size less than the wavelength of visible light, giving the unique optical properties.

4. Production of carbon soot (AS and MS)/PMMA nanocomposites

The synthesizing technique in the production of carbon soot polymer nanocomposites was a mixture of solution mixing and solvent casting. In this section, synthesized nanocomposites for two different nanofillers are discussed: one is AS/PMMA nanocomposite and the other one is MS/PMMA nanocomposites.

- For the solution of PMMA, PMMA was dissolved in 20 ml dichloromethane (DCM) solvent using the help of magnetic stirrer for a duration of 3 h (**Figure 13**).

- Nanofiller carbon soots (almond soot particles and mustard soot particles) 1 and 2 mg were dispersed in the polymer solution by ultrasonication for 4 h

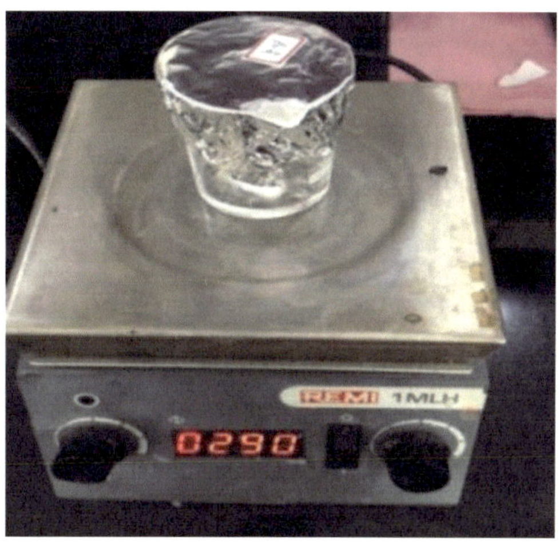

Figure 13.
Actual setup of pure PMMA solution by magnetic stirrer and ultrasonication.

Figure 14.
Actual setup of CS/PMMA homogeneous solution by magnetic stirrer and ultrasonication.

(at the end temperature of sonication was 35°C) to get a homogeneous solution (**Figure 14**).

- At the end, a homogeneous solution was purged in a flat glass Petri dish, which was floating over mercury in vacuum box.

- We take the prepared homogeneous solution for approximately 24 hr in a Petri dish in low vacuum (8-10mTorr) at room temperature, and after the thin film samples reach a dry state, they are removed from Petri dishes.

- Thin film of polymer (PMMA) without the nanofiller (carbon soots) was also prepared for the reference composite used in the comparison study of many composite properties. The film of pure PMMA was transparent and other composite films were opaque and in black color due to dispersion of carbon soots. All produced films were in the thickness of 100 μm for pure PMMA and different concentration (1 and 2 mg) of carbon soot nanofillers.

5. Properties of CS/PMMA nanocomposites (CSPNCs)

5.1 Surface morphological and chemical composition properties

Atomic force microscopy study is a characteristic and quantitative analysis for measuring the surface roughness at nanodimension and visualizing the nanotexture of the deposited thin film surface. The topography deflection image and 3D AFM images of the pure PMMA and CS/PMMA nanocomposites, in which the roughness value of nanocomposites are carried out, are shown in **Figure 15a–c**.

The roughness parameter with average roughness (R_a), root mean square roughness (R_q), skewness of the line (R_{sk}) and kurtosis of the line (R_{ku}) roughness data are shown in **Table 3**. The pure PMMA thin film, AS/PMMA and MS/PMMA thin films are synthesized for comparative analysis of surface roughness. The measures of roughness of carbon soot polymer thin films are less than the roughness of pure PMMA film. This implies that the prepared PMMA thin films with almond and mustard soot decrease the surface roughness. The ration of roughness value R_q and R_a (R_q/R_a) are approximately close to theoretical data, recorded by Ward et al. in 1982 [42].

The chemical composition for CSPNCs was carried out by EDAX (energy dispersive X-ray analysis). Results from EDAX depicted in **Figure 16a–c** reveal the chemical composition as follows:

Carbon approximately 75% and remaining oxygen approximately 25% for all nanocomposite samples with individual content of PMMA, AS/PMMA and MS/PMMA are 76.31, 76.20, 75.26% and 23.69, 23.80, 24.74% respectively (**Table 4**). These CS (AS and MS)/PMMA samples are electrically nonconducting, and for EDAX analysis, these samples are coated with gold to convert them into electrically conducting. Due to this, an extra small peak is observed for gold at approximately 2 keV in EDAX spectra.

5.2 Spectroscopic properties

The FTIR spectra of PMMA show sharp intense peaks at 1725 and 1141 cm^{-1} and are attributed to the ester carbonyl stretching and C–O–C bending vibration. In addition, peaks appeared at 3000 and 2950 cm^{-1} and are assigned to the C–H stretching vibrations, while the peaks at 752 and 840 cm^{-1} are attributed to the

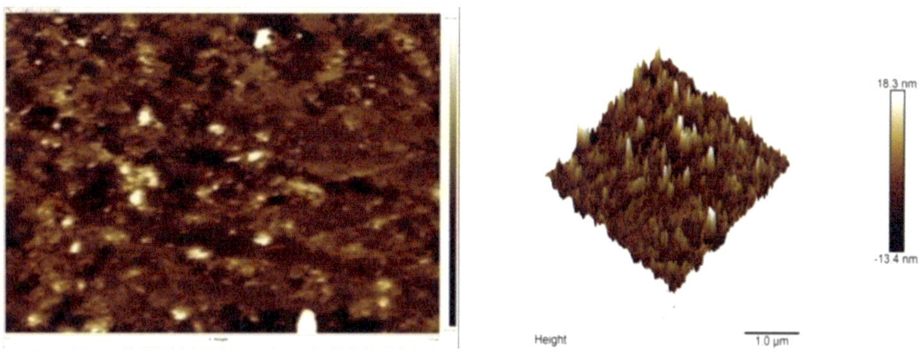

(a)

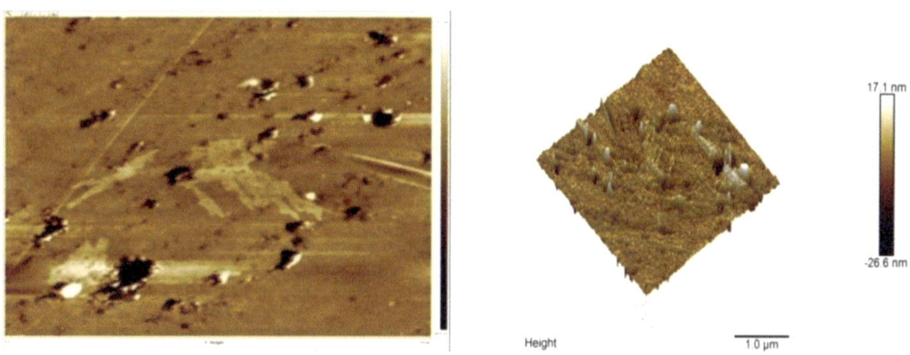

(b)

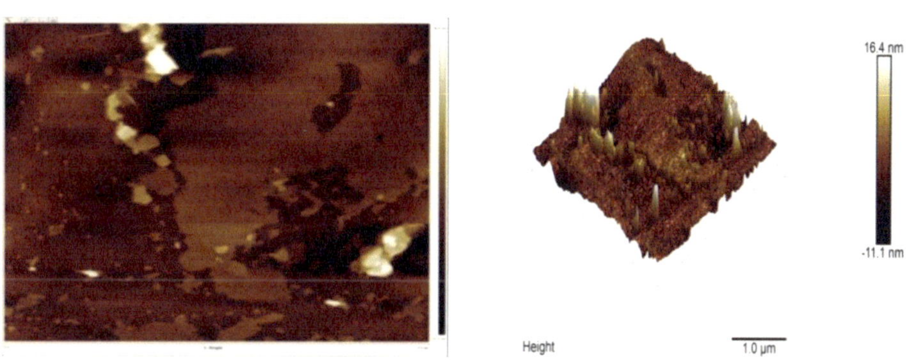

(c)

Figure 15.
Topographic deflection image and 3D image of (a) pure PMMA, (b) AS/PMMA, and (c) MS/PMMA thin film.

vibration of polymethyl methacrylate chains and the deformation vibration of PMMA O—C—O, respectively. Moreover, the peaks at 1236, 1383 and 1440 cm^{-1} were attributed to the ester band symmetrical and asymmetrical stretching vibrations. The peak at 984 cm^{-1} represented C—H bending wagging vibration. It should be noted that there is no peak near 1680 to 1640 cm^{-1} range, which confirms that all methyl methacrylate (MMA) monomers are converted into PMMA polymer. The FTIR patterns of almond soot-doped PMMA viz. pure and mustard soot-doped PMMA viz. pure are shown in **Figures 17** and **18**, respectively.

Samples	Roughness parameters				
	R_a (nm)	R_q (nm)	R_q/R_a	R_{sk}	R_{ku}
Pure PMMA	2.99	4.40	1.47	2.64	25.5
AS/PMMA	2.18	4.10	1.88	−3.36	36.1
MS/PMMA	2.05	3.10	1.51	2.02	15.2

Table 3.
The roughness parameters of CS/PMMA nanocomposites.

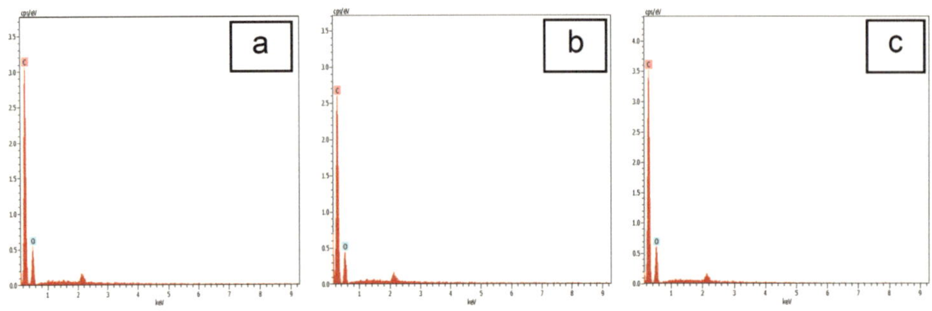

Figure 16.
EDAX spectra of (a) PMMA, (b) almond soot/PMMA, and (c) mustard soot/PMMA samples.

The doping level of almond soot and mustard soot fillers is 1 and 2 mg, respectively. Two broad peaks are centered at 1725 and 1141 cm^{-1} in AS/PMMA and MS/PMMA.

The other peaks are at 3000, 2950, 1440, 1383, 1236, 984, 839, 752 and 481 cm^{-1}, which correspond to the formation of PMMA matrix. If we compare the pure PMMA and AS/PMMA or MS/PMMA nanocomposites, we will find there is no change in wave number (**Figures 17** and **18**). These show that all AS and MS filler particles were loaded inside the polymer matrix without changing their chemical structure with low intensity as the AS and MS particle concentration increases.

5.3 Optical properties

In this section, we discuss the optical properties of polymethyl methacrylate (PMMA), almond soot/PMMA and mustard soot/PMMA nanocomposites for different concentration. The absorbance spectra of nanocomposite thin films were recorded in the range between 200 and 800 nm.

5.3.1 Absorption data

The recorded absorbance spectra of pure PMMA and almond soot (**Figure 19a**) and pure PMMA and mustard soot (**Figure 19b**) nanocomposites with different concentration at the wavelength of 200–800 nm show that the absorbance value increases with the addition of increased carbon soot (AS/MS) nanofillers.

As the absorption peaks increase, these show the existence of internal chemical interaction between the carbon soot nanofillers and PMMA matrix. In this manner, our study of optical properties for PMMA and doped freshly carbon soots such as almond soot particles and mustard soot particles at a different concentration give us good results. The result from synthesized carbon aggregates and aerosol particles

Samples	Elemental compositions	
	C%	O%
Pure PMMA	76.31	23.69
AS/PMMA	76.20	23.80
MS/PMMA	75.26	24.74

Table 4.
Elemental compositions of PMMA, AS/PMMA and MS/PMMA.

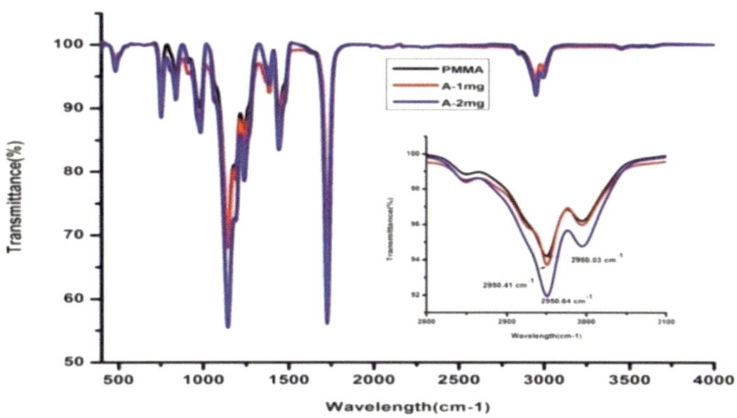

Figure 17.
The FTIR spectra of pure PMMA and AS/PMMA nanocomposites.

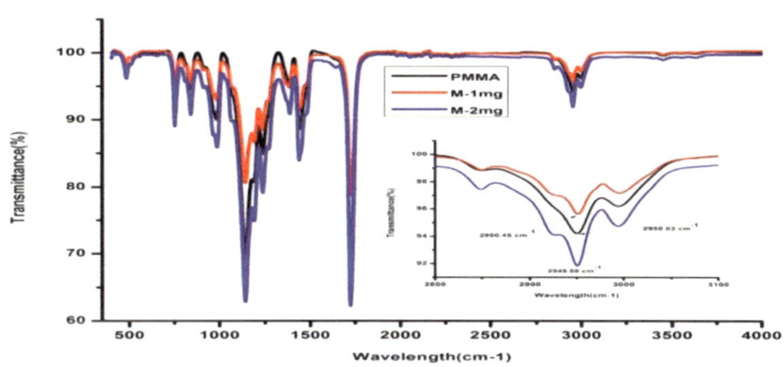

Figure 18.
The FTIR spectra of pure PMMA and MS/PMMA nanocomposites.

investigated by light scattering and light absorption shows a good agreement with results obtained by other research groups [43, 44].

5.3.2 Absorption coefficient data analysis

The absorbance coefficients of PMMA, AS/PMMA (**Figure 20a**) and PMMA, MS/PMMA (**Figure 20b**) nanocomposite thin films are shown in the below figure.

According to a review, if the value of α is high or $\alpha > 10^4\ cm^{-1}$, then the electron direct transition is expected. If the value of α is low or $\alpha < 10^4\ cm^{-1}$, then the expected transition of electron is indirect. In our study, the absorption coefficient (α) values of the polymer matrix (PMMA) and all filler nanocomposites

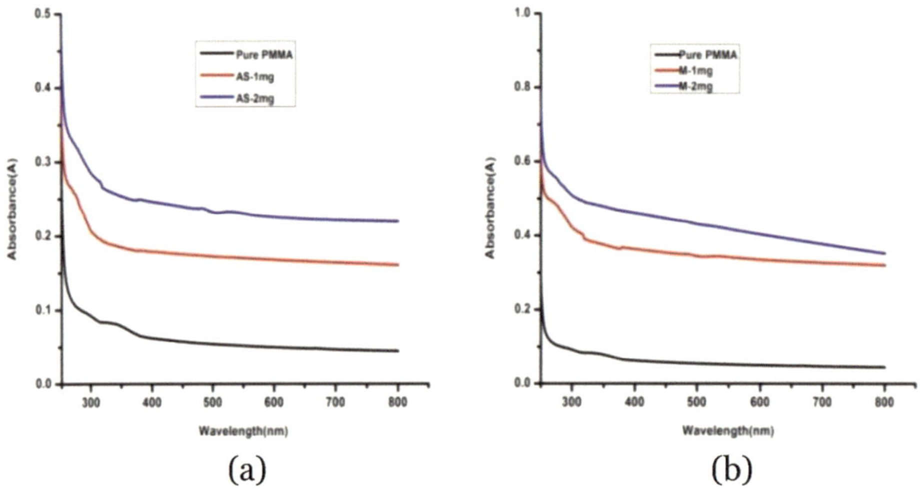

Figure 19.
Absorbance spectra of pure PMMA with (a) AS/PMMA nanocomposites and (b) MS/PMMA nanocomposites.

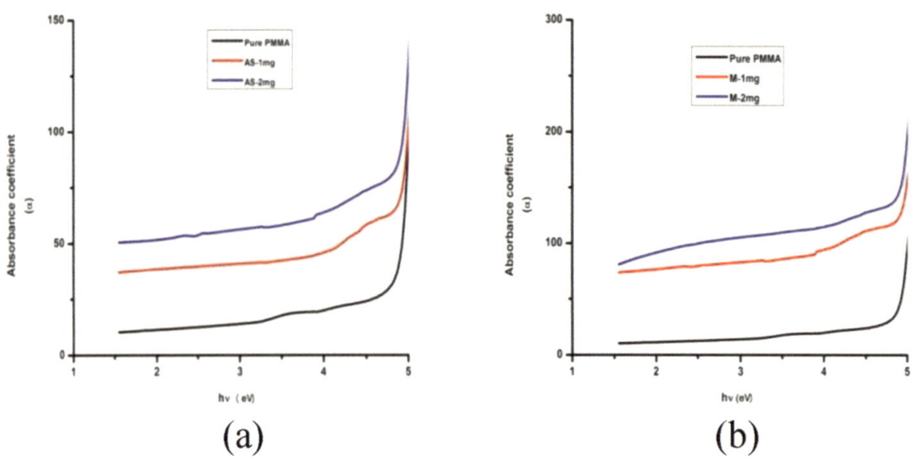

Figure 20.
Variation of absorption coefficient of PMMA with (a) AS-doped PMMA nanocomposite thin films and (b) MS-doped PMMA nanocomposite thin films.

are not >10^4 cm^{-1}, and this implies the indirect electron transition in AS and MS conditions.

In this study of nanocomposite optical properties, we used carbon nanosoot obtained from incomplete combustion of flame deposition method similar to "Sara D." et al. in 2018. They found the absorption coefficient for two different flame-generated soot particles, in which the results for both cases were identical [45]. The results of absorbance coefficient for PMMA matrix and carbon soots-doped PMMA nanocomposites indicate indirect electron transition.

5.3.3 Band gap analysis

The band gap plots of polymer matrix and AS/PMMA (**Figure 21a**) and polymer matrix and MS/PMMA (**Figure 21b**) nanocomposites on the variation of (αhν) 1/2 and (hν) are represented. These band gap spectra show that the measurements

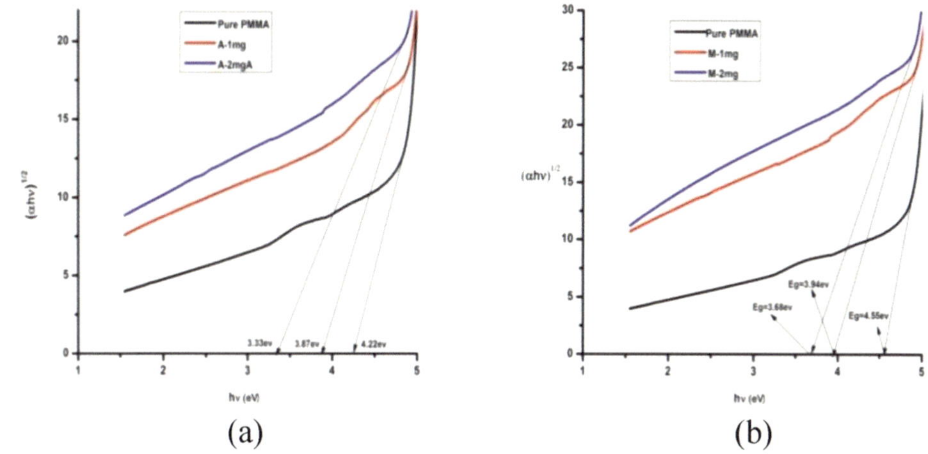

Figure 21.
The band gap spectra of PMMA with (a) AS/PMMA thin films and (b) MS/PMMA thin films.

Samples	Energy band gap values E_g (eV)
Pure PMMA for AS	4.22
AS-1 mg/PMMA	3.87
AS-2 mg/PMMA	3.68
Pure PMMA for MS	4.55
MS-1 mg/PMMA	3.94
MS-2 mg/PMMA	3.68

Table 5.
Energy band gap values of PMMA, almond soot and mustard soot nanocomposites.

revealed that the concentration of filler nanoparticles affects the optical properties of PMMA matrix. The decreasing values of optical energy band gap with adding the carbon soot in PMMA, are located in **Table 5**.

6. Production of carbon soot (AS and MS)/PS nanocomposites

In this study, we have prepared the polymeric thin film of host matrix PS and carbon soot-doped PS nanocomposites at a different weight percentage of almond and mustard soot of 50 μm thickness by solution casting technique.

The steps for preparation of CS/PS nanocomposite thin films are similar as discussed above in Section 4.

7. Properties of CS/PS nanocomposites (CSPNCs)

7.1 Surface morphological and chemical composition properties

We measure the surface topology and roughness of CS/PS composites with horizontal length scale in micrometer and vertical scale in the range of nanometer (**Figure 22a–c**).

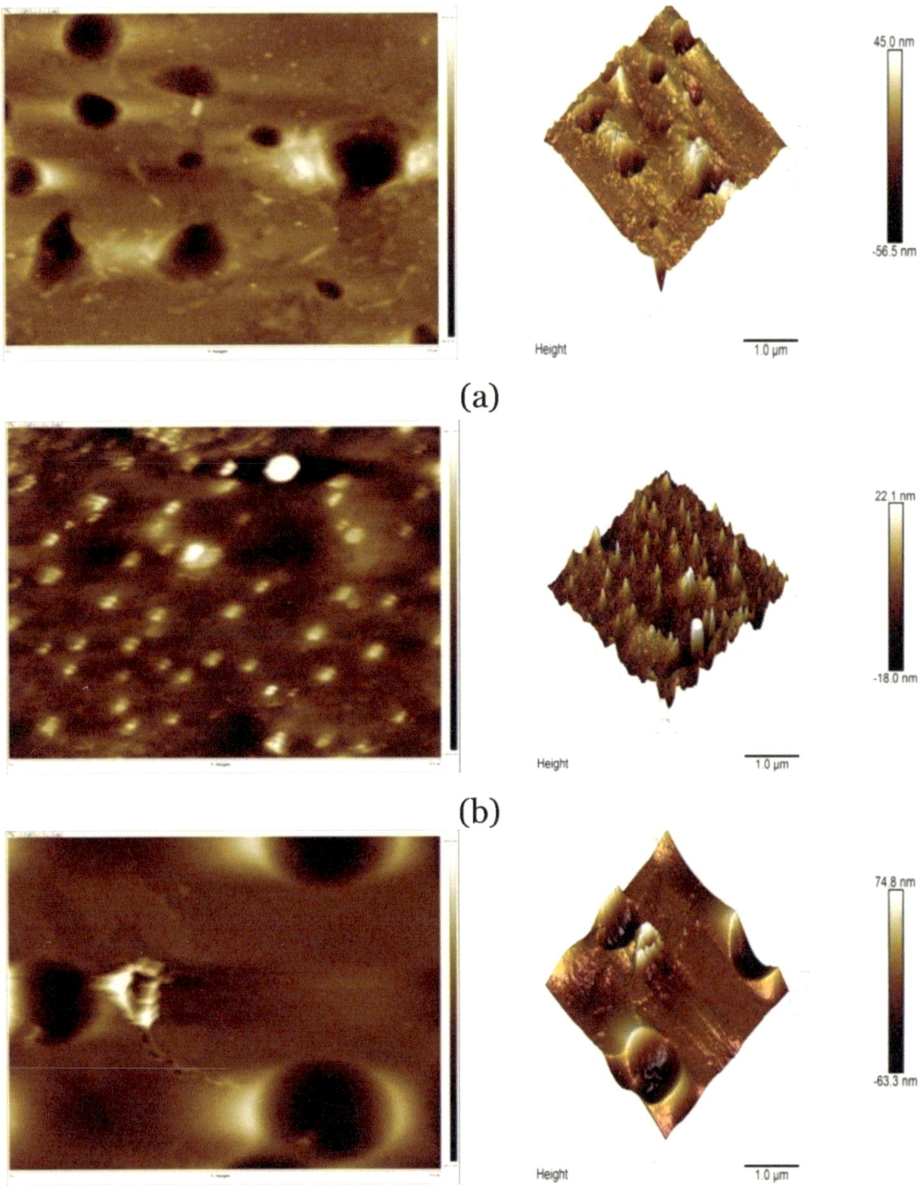

Figure 22.
Topographic deflection image and 3D AFM image of (a) PS (pure polystyrene), (b) AS/PS nanocomposite thin film and (c) MS/PS nanocomposite thin film.

The roughness parameters of carbon soot polymer nanocomposites (CSPNCs) including the average surface roughness (R_a), root mean square roughness (R_q), skewness of the line (R_{sk}) and kurtosis of the line (R_{ku}) roughness are obtained by AFM machine. The average surface roughness for pure PS, AS/PS and MS/PS with values of 2.99, 2.18 and 2.05 nm is carried out as well as the root mean square roughness for pure PS, AS/PS and MS/PS nanocomposites with the values of 4.40, 4.10 and 3.10 nm is obtained. Also, R_{sk} and R_{ku} values of all samples are found out, while the rations of R_q/R_a of all samples are calculated for all composites, as shown in **Table 6**.

As mentioned in the table, the surface roughness decreases with the doping of almond soot in PS matrix and the surface roughness increases with the doping of mustard soot in PS matrix, as compared with the pure polystyrene sample. The obtained result reveals that surface of AS/PS nanocomposites is smoother as compares to that of MS/PS and pure PS sample. We calculated the ratio of root mean square roughness and average roughness, and the ratio of R_q/R_a almost matched the value of theoretical ratio data. The surface roughness of composite samples plays an important role in the wetting properties.

The analytical composition of soot filler nanocomposites carried out by energy dispersive X-ray spectroscopy of studied samples is represented in **Figure 23a–c**. We can see that all samples contained approximately 98–99% carbon and remaining 1–2% oxygen element (**Table 7**).

It is confirmed that the sample composed of only carbon and oxygen element, without any other element. These CS (AS and MS)/PS samples are electrically non-conducting, and for EDAX analysis, these samples are coated with gold to convert them into electrically conducting. Due to of this, an extra small peak is observed for gold at approximately 2 keV in EDAX spectra (**Figure 23**).

7.2 X-ray diffraction properties

The XRD spectra of pure polystyrene almond soot/PS (**Figure 24a**) and mustard soot/PS (**Figure 24b**) nanocomposites at different weight concentration (1 and 2 wt%) are in the range of 2θ = 20–90°. The X-ray diffraction pattern from pure PS, AS/PS nanocomposites with 1 and 2 wt% and MS/PS nanocomposites with 1 and 2 wt% concentrations does not show any visible peaks. This corresponds to the pure amorphous polymeric structure without any peak and suggested that the carbon soot particles had been exfoliated in the soot nanocomposites, as referred in [46, 47]. As shown in **Figure 24a** and **b**, the XRD pattern of AS/PS and MS/PS nanocomposites at different weight concentrations does not loss own nature. All results suggest that the variation of intensity confirms the presence of carbon soot in polymer nanocomposites.

7.3 Spectroscopic properties

The FTIR pattern of pure polystyrene shown as a broad intense peak at 694 cm^{-1} is attributed to the (C–H) bend due to the ring deformation vibration. A peak at 2918 cm^{-1} is attributed to C–H stretching with asymmetric CH_3 group and C=O stretching around 1739 cm^{-1}.

The two peaks at 1600 and 1491cm^{-1} are assigned due to aromatic C=C stretching, and the other two peaks at 1450 and 1366 cm^{-1} are attributed to the ester carbonyl stretching vibration and C–H bending vibration. Some bands at 1216, 1027 and 905 cm^{-1} in the range 600–1200 cm^{-1} correspond to the plane deformation of

Samples	Surface roughness parameters				
	R_a (nm)	R_q (nm)	R_q/R_a	R_{sk}	R_{ku}
Pure PS	6.69	11.0	1.64	−1.22	8.44
AS/PS	3.53	5.50	1.55	2.61	25.6
MS/PS	10.3	16.3	1.58	0.65	6.67

Table 6.
The surface roughness parameters of carbon soot/polystyrene nanocomposites.

Samples	Percentile compositions	
	C%	O%
Pure PS	99.36	0.64
AS/PS	99.10	0.90
MS/PS	98.87	1.13

Table 7.
Percentage of element compositions of PS, AS/PS and MS/PS thin film.

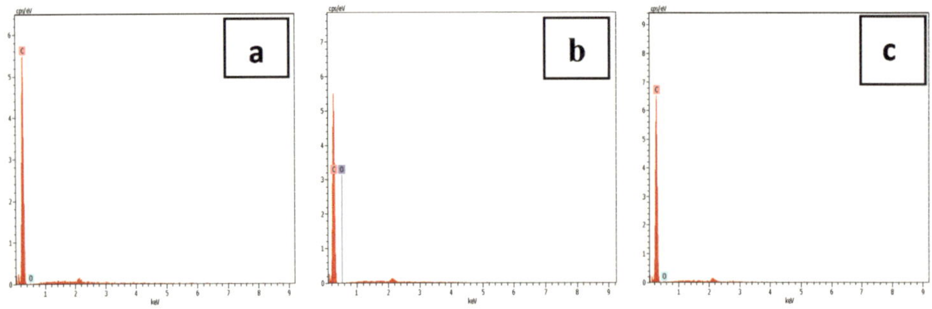

Figure 23.
EDAX spectra of (a) PMMA, (b) almond soot/PMMA and (c) mustard soot/PMMA samples.

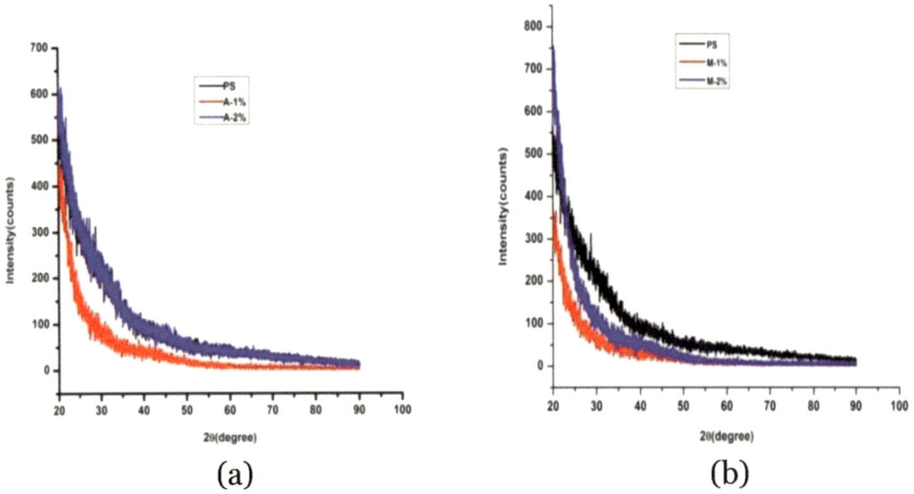

Figure 24.
XRD spectra of (a) almond soot/PS nanocomposites and (b) mustard soot/PS nanocomposites.

C–H group bending. Another peak at 750 cm^{-1} shows due to (C–H) deformation vibration band of benzene ring hydrogen.

The comparative study on Fourier transform infrared spectra of pure polystyrene vs doped carbon soot/PS nanocomposites shows (**Figures 25** and **26**) the peaks around 2920, 1739, 1600, 1492, 1451, 1367, 1215, 1027, 905, 750 and 537 cm^{-1} corresponding to the formation of PS. Bands of CS/PS nanocomposites are more and less as compared to the peaks of pure polystyrene, but there are no other peaks found, showing that all the soot nanoparticles are loaded inside the PS host matrix without changing their chemical structure. No other modification in the shift of chemicals and the shape of band is noticed in the carbon soot/polystyrene nanocomposite spectra, indicating no chemical modification occurred in the polystyrene.

7.4 Optical properties

In this section, we discuss about the optical properties of polystyrene nanocomposites. Here, the carbon soot was doped with either almond soot or mustard soot particles with 1 and 2 wt%, respectively. In the optical properties, we tried to understand the absorption coefficient properties. Therefore, we measured the band gap of nanocomposites in the range of 200–800 nm.

7.4.1 The absorption data

The absorbance data spectra of pure polystyrene and AS/PS (**Figure 27a**) and pure polystyrene and MS/PS (**Figure 27b**) nanocomposite thin films show the value of absorbance increases with doping the carbon soot in PS as compared to the pure PS. The increased absorbance in spectral profile from nanocomposite reveals the correlation between carbon soot nanoparticles and the host matrix polystyrene matrix. In this study, carbon soot (traditional Indian name—"Kajal") is produced by incomplete combustion of almond and mustard oil by flame deposition method. Researchers also used different types of soot from aerosol carbon particles, diesel soot and other sources. Similar results were found by Mita et al. in 1980 [48] from absorbance of aerosol soot particles. As seen in the figures, the value of absorbance is low at high wavelength and high absorbance value corresponding to the low wavelengths.

7.4.2 Absorption coefficient and band gap study

The absorption coefficient spectra of polystyrene and soot/polymer nanocomposites are represented in **Figure 28a** and **b**.

For the indirect band gap, the value of r is 1/2; hence, the indirect band gap of nanocomposite thin films is obtained using the help of plot $(\alpha h\nu)^{1/2}$ Vs $h\nu$ (**Figure 29a** and **b**). It has been noticed that the band gap is inversely proportional

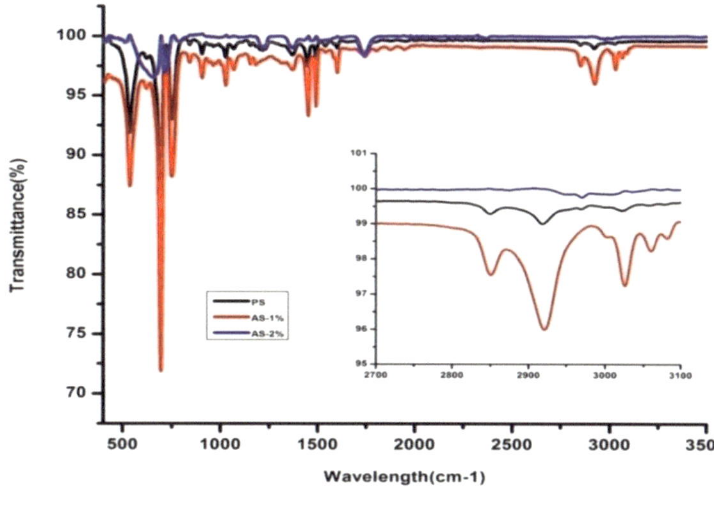

Figure 25.
FTIR spectra of almond soot polystyrene nanocomposites.

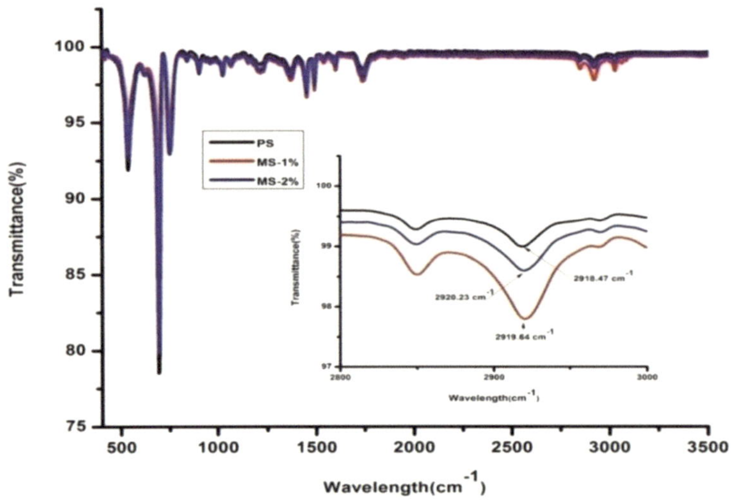

Figure 26.
FTIR spectra of mustard soot polystyrene nanocomposites.

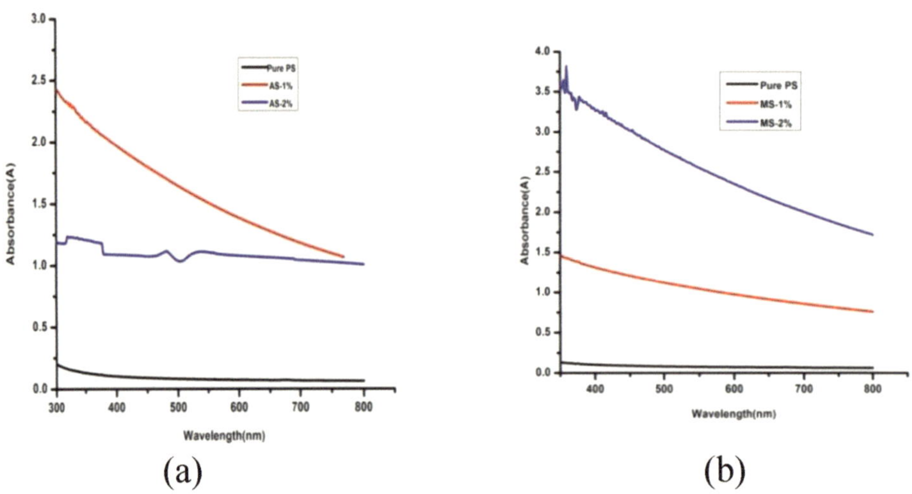

Figure 27.
The absorbance data spectra of pure polystyrene with (a) almond soot-doped PS thin films and (b) mustard soot-doped PS thin films.

to the thickness of films. The indirect band gap of pure PS and AS/PS at 1 and 2 wt% is 4.06 eV, 3.84 and 3.47 eV, as well as for PS, MS/PS at 1 wt% and MS/PS at 2 wt%, it is 4.33, 3.94 and 3.85 eV. Energy band gap values for all thin film samples of 50 μm thickness are represented in **Table 8**.

7.5 Glass transition temperature phenomenon (T_g)

Dynamic mechanical analysis is an amazing tool to study the viscoelastic properties of carbon soot particles filled with polymer. The other authors investigated the mechanical properties of carbon nanoparticles with polymer [48–53]. The dynamical mechanical results of carbon soot/polystyrene nanocomposites produced by solution casting method for pure polystyrene, almond soot/polystyrene and mustard soot/polystyrene at 1 and 2 wt% concentrations (**Figure 30**) show that for all CS/PS nanocomposites there is a downshift in viscous modulus toward a higher temperature compared to the pure polystyrene. All data were recorded at 1 Hz frequency and 22–160°C temperature.

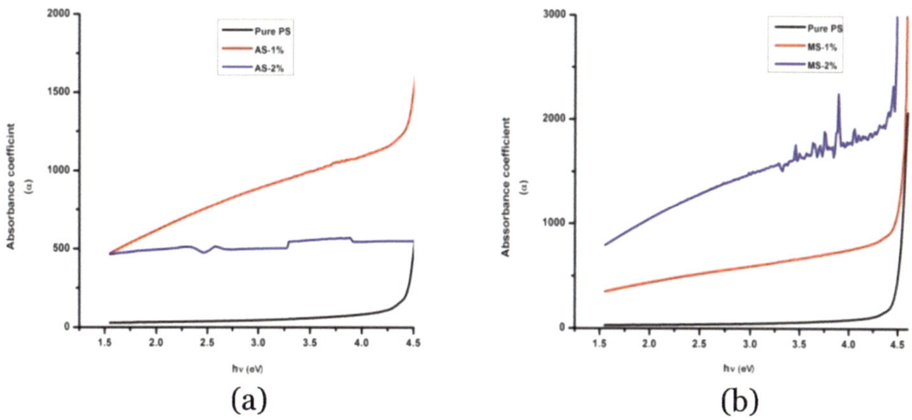

Figure 28.
Representation of absorption coefficient spectra of pure polystyrene with (a) AS/PS composite for 1 and 2 wt% concentration and (b) MS/PS composite for 1 and 2 wt% concentration.

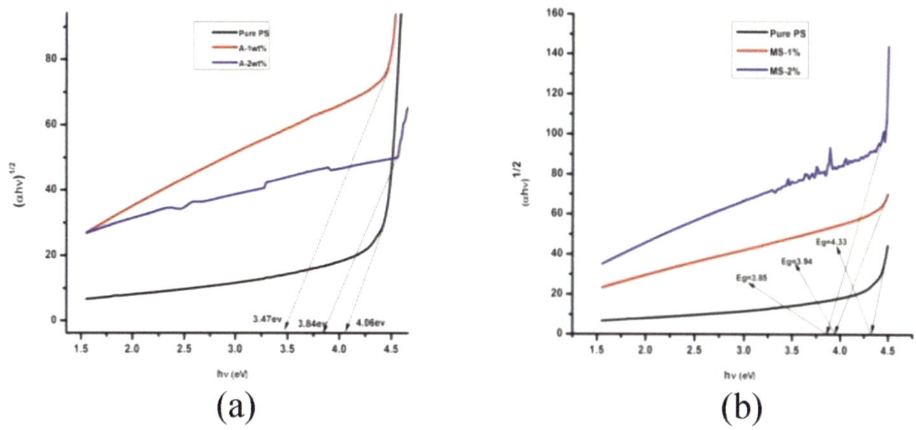

Figure 29.
Plot of (αhν) 1/2 vs hν photon energy (hν) for the pure polystyrene with (a) almond soot 1 and 2 wt% in PS and (b) mustard soot 1 and 2 wt% in PS.

PS and CSPNC samples	Energy band gap E_g (eV)
Pure PS for AS samples	4.06
AS-1 wt%	3.84
AS-2 wt%	3.47
Pure PS for MS samples	4.33
MS-1 wt%	3.94
MS-2 wt%	3.85

Table 8.
The obtained value of pure polystyrene energy band gap and AS/PS and MS/PS at 1 and 2 wt% concentration.

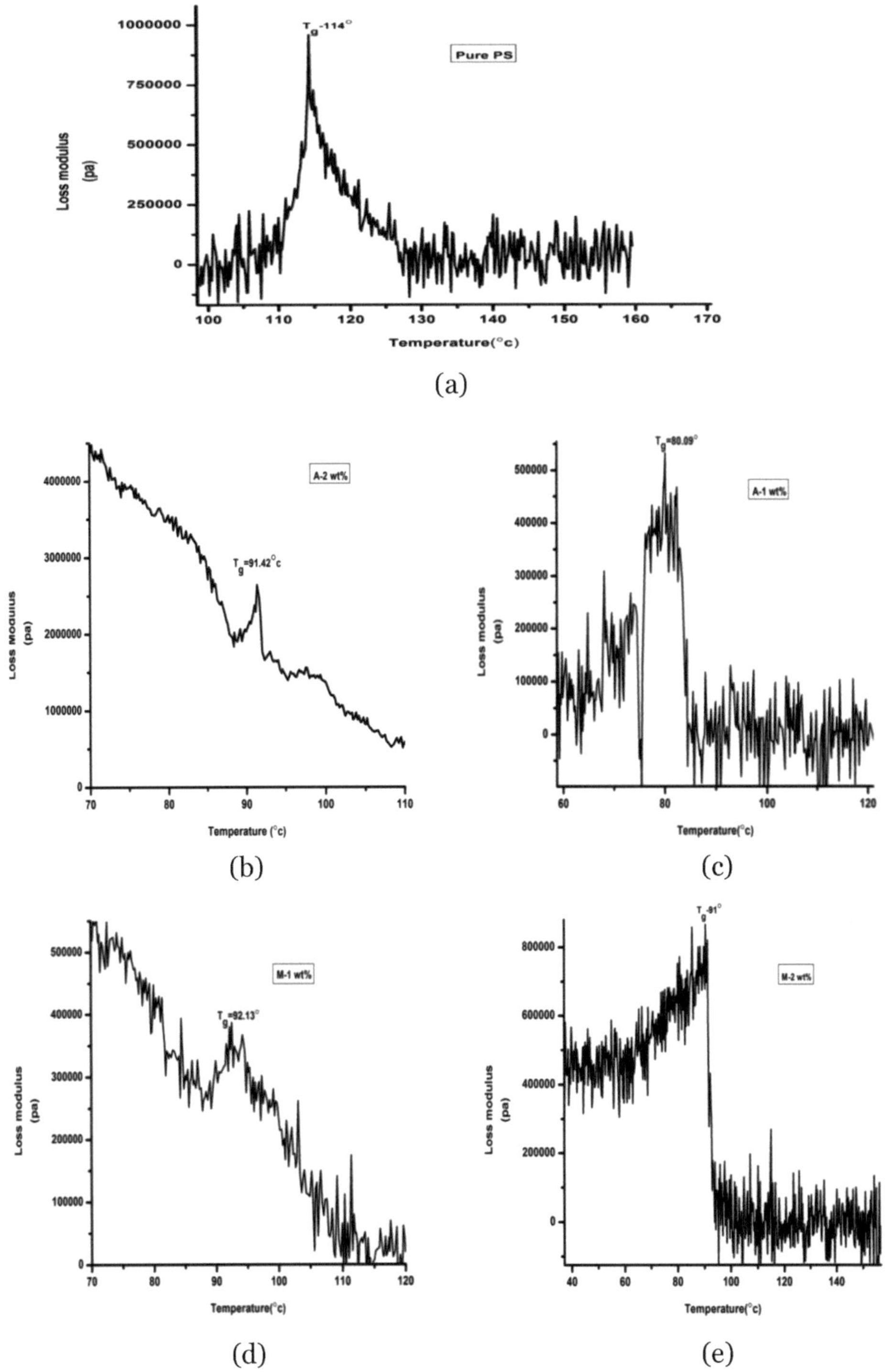

Figure 30.
Variations of viscous modules (E") with temperature for (a) pure polystyrene, (b) almond-1 wt%/PS, (c) almond-2 wt%/PS, (d) mustard-1 wt% PS and (e) mustard-2 wt%/PS.

The glass transition temperature (T_g) of pure polystyrene is 114°C, and glass transition temperatures for almond soot/PS nanocomposites for 1 and 2 wt% are 80 and 91°C, as well as for mustard soot/PS nanocomposites at 1 and 2 wt% concentration are 92 and 91°C (**Table 9**). We can see that the value of glass transition

Samples	Glass transition temperature T_g (°C)
Pure PS	114
A-1 wt%	80
A-2 wt%	91
M-1 wt%	92
M-2 wt%	91

Table 9.
The glass transition temperature (T_g) values of CS/PS nanocomposites.

temperature decreases for all samples as compared to the pure polystyrene. In the case of AS/PS nanocomposites, the value of T_g increases with the concentration of almond soot nanoparticles and reveals that the addition of more soot particles hinders the segmental relaxation of the polystyrene chains, while for the MS/PS nanocomposites, the value of T_g slightly decreases with the increase of mustard soot nanoparticles.

8. Conclusions and future scopes

8.1 Conclusions

The chapter is concerned with two things: first with carbon soot nanoparticles (CSNPs) and second is carbon soot polymer nanocomposites (CSPNCs). The morphological and spectroscopic analysis is carried out for carbon soot nanoparticles, and the glass transition temperature phenomenon, and optical and structural properties are obtained for the synthesized carbon soot polymer nanocomposites.

Production of carbon soot from incomplete combustion of almond oil and mustard oil by low cost flame deposition method in two different environment one is natural environment (without water tub) and other one is with water tub, these prepared fresh carbon soots such as almond soot (AS) and mustard soot (MS) are characterized by FESEM, XRD, EDAX, FTIR and UV for morphological, elemental composition and spectroscopic analysis. The formed carbon soot particles are the amorphous nanomaterial with the size of approximately 50 nm, confirmed by the XRD and FESEM study. The IR study is carried out to identify the presence of any functional group, some small SP^2, SP^3 and aromatic clusters. Here, with water tub condition is an important tool for identifying the morphology and particle size of carbon nanoparticles (CNPs). The synthesized soot is useful for die purpose, paint, marker pen ink, etc. It is also used for pigment and reinforcement in vehicle tires as it decreases the thermal damage.

In the next step, we produced the carbon soot polymer nanocomposites (CSPNCs) for different carbon soot nanoparticles such as almond soot and mustard soot with different polymer matrices of polymethyl methacrylate (PMMA) and polystyrene (PS).

Primarily, we synthesized carbon soot/PMMA nanocomposites with different concentration 1 and 2 mg of AS and MS of 100 μm thickness by solution casting technique. Prepared samples are characterized for the optical and surface morphological properties. FESEM and EDAX provide the useful information about the nanostructure and compositions of nanocomposite samples. FTIR analysis of CS/PMMA nanocomposites shows that the soot nanoparticles are loaded in polymer (PMMA) without change in chemical structure for the surface morphology calculating the surface roughness of all samples, and the decreasing values of surface roughness with the doping of soot nanoparticles in host matrix PMMA are obtained. In this sequence, the band gap of PMMA and all CS/PMMA nanocomposites are calculated for the optical properties. The lower values of energy band gap are carried out by filling CS in PMMA as compared to the pure PMMA.

For the second case of carbon soot polymer nanocomposites, carbon soot/PS nanocomposites were produced with 1 and 2 wt% concentration of AS and MS nanofillers in host matrix polystyrene by cost-effective solution cast method of 50 μm thickness. The glass transition temperature phenomenon, optical, structural and surface morphological studies are carried out for prepared composite samples. X-ray diffraction study reveals the variation of intensity showing the presence of carbon soot in the polystyrene chain and gives the important information about the amorphous nature. The decreasing values of roughness and band gap are obtained with the doping of carbon soot nanoparticles in the host polymer matrix polystyrene, and also the low values of glass transition temperature T_g of CS/PS samples are obtained as compared to the pure PS sample. These obtained properties of carbon soot polymer nanocomposites make them very useful in various applications including in the separation of oil and water, anti-icing and energy-saving buildings, high-energy vehicles, electric trains and electric devices.

8.2 Future scope

The incomplete combustion synthesis of carbon soot is a relatively new field of research that has not yet been exploited in all its potentiality. A good controlled incomplete combustion process is flame deposition has represented to be able to synthesis of different carbonaceous particles from burning different oils in a low temperature range. This study provide a deep knowledge of the soot species production method under a flame environment. It was carried out on the experimental results basis. A good relationship between production/combustion conditions and carbon properties like morphological, chemical composition, spectroscopic and internal structure has been found. We can produce the soot by any other method.

Nevertheless, the main focus on investigation of carbon allotropes such as graphine, carbon nanotube (CNT), fullerenes etc by flame deposition method for more demanding applications. Ultimately, the synthesized soot particles through combustion are suitable for industrial applications.

Carbon soot nanoparticles in PMMA and PS could be improved if these nanoparticles are functionalized by some surfactants, and this will be a more homogeneous dispersion of nanoparticles in polymers. Thin films of CS/PMMA or PS could also be casted by spin coating technique. Mechanical and optical properties of CS/polymer samples give faithful results; we can investigate the di-electric and thermal conductivity for further industrial applications.

Conflict of interest

The authors declare no conflict of interest.

Abbreviations

AFM	atomic force microscopy
AS	almond soot
CNPs	carbon nanoparticles
CNTs	carbon nanotubes

CS	carbon soot
CSPNCs	carbon soot polymer nanocomposites
DMA	dynamic mechanical analyzer
EDAX	energy dispersive X-ray
FESEM	field emission scanning electron microscopy
FTIR	Fourier transform infrared spectroscopy
PMMA	polymethyl methacrylate
PS	polystyrene
UV	ultraviolet
XRD	X-ray diffraction

Author details

Rakhi Tailor*, Yogesh Kumar Vijay and Minal Bafna
Vivekananda Global University, Jaipur, India

*Address all correspondence to: rrtrbt@yahoo.com

References

[1] Whelan P. The mechanics of the formation and adhesion of deposits arising from the combustion of liquid fuels. The Institution of Mechanical Engineers. 1961:1-17

[2] Glassman I. Combustion. New York (London): Academic Press; 1977. p. 242

[3] Grisdale R. The formation of black carbon. Journal of Applied Physics. 1953;**24**:1082

[4] Koylu U, Faeth G, Farias TL, Carovalho M. Fractual and projected structure properties of soot aggregates. Combustion and Flame. 1995;**100**:621-633

[5] Jensen K, Suo-Antilla J, Blevins L. Characterization of soot properties in two meter JB-8 pool fires. In: Technical Report 2005-0337. California: Sandia National Laboratories; 2005

[6] Eliot R. Boiler Fuel Additives for Pollution Reduction and Energy Saving. New Jersey: Noyes Data Corporation; 1978. p. 8

[7] Lindsey A. The formation of polycyclic aromatic hydrocarbons carbon deposits from normal and reversed diffusion flames. Combustion and Flame. 1960;**4**:261-264

[8] Seinfield J, Pandis S. Atmospheric Chemistry and Physics: From Air Pollution to Climate Change. New York: Wiley Interscience; 1997

[9] Lynn M. Soot diagnostics based on leaser heating. Applied Optics. 1984;**23**:2201-2208

[10] Joyce D. The life cycle of a debris particle. Tribology and Interface Engineering Series. 2005;**48**:681-690

[11] Rooca L, Liberto G, Shayler P, Fay M. The nanostructure of soot in oil particles and agglomerates from an automotive diesel engine. Tribology International. 2013;**61**:80-87

[12] Tami C, Robert W. Light absorption by carbonaceous particles: An investigative review. Aerosol Science and Technology. 2006;**40**:27-67

[13] Rocca L, Liberto G, Shayler P. Application of nanoparticle tracking analysis platform for the measurement of soot in oil agglomerates from automative engines. Tribology International. 2014;**70**:142-147

[14] Growney D, Mykhaylyk O, Middlemiss L. Is carbon black a suitable model colloidal substrate for diesel soot? Langmuir. 2015;**31**:10358-10369

[15] Saito K, Gordon A, Williams F, Stickle W. A study of early history of soot formation in various hydrocarbon diffusion flames. Combustion Science and Technology. 1991;**80**:103-119

[16] Megaridis C, Dobbins R. Comparison of soot growth and oxidation in smoking and non-smoking ethylene diffusion flames. Combustion Science and Technology. 1989;**66**:1-16

[17] Chen J, Castangnoli C, Niksa S. Coal devolatilization during rapid transient heating. Energy and Fuels. 1992;**6**:264-264

[18] Haynes B, Wagner H. Soot formation progress. Progress in Energy and Combustion Science. 1981;**7**:229-273

[19] Fitzer E, Kocling K, Boetim H, Marsh H. Recommended terminology for the description of carbon as a solid. Pure and Applied Chemistry. 1995;**67**:473-506

[20] Bruce C, Stromburg T, Gurton K, Mozer J. Trans spectral absorption

and scattering of electromagnetic radiation by diesel soot. Applied Optics. 1991;**30**:1537-1546

[21] Desantes J, Bermudez V, Garcia J, Fuentes E. Effects of current engine strategies on the exhaust aerosol particle size distribution from a heavy duty diesel engine. Journal of Aerosol Science. 2005;**36**:1251-1276

[22] Vanderwal R, Tomasek A, Street K, Hull D, Thompson W, Sljenbug J. Carbon nanostructure examined by lattice fringe analysis of high resolution transmission electron microscopy. Applied Spectroscopy. 2004;**58**:230-237

[23] Vandernial R, Tomasek A. Soot oxidation: Dependence upon initial nanostructure. Combustion and Flame. 2003;**134**:1-9

[24] Amann C, Siegla D. Diesel particulates—What they are and why. Aerosol Science and Technology. 1982;**1**:73-101

[25] Esangbedo C, Boehman A, Peres J. Characteristics of diesel engine soot that lead to excessive oil thickening. Tribology International. 2011;**47**:194-203

[26] Clague A, Donnet J, Wang T, Peng J. A comparison of diesel engine soot with carbon black. Carbon. 1999;**37**:1553-1565

[27] Stanmore B, Brilhac J, Gilot P. The oxidation of soot: A review of experiments, mechanism and models. Carbon. 2001;**39**:2247-2268

[28] Sarotim A, Longwell J, Wornat M, Mukherjee J. The role of biaryl reactions in PAH and soot formation. In: Soot Formation in Combustion. Springer: Series in Chemical Physics, CHEMICAL; 1994;**59**:485-499

[29] Tree D, Svensson K. Soot processes in compression ignition engines. Progress in Energy and Combustion Science. 2007;**33**:272-309

[30] Gaydon A, Wolfhard H. Flames. London: Chapman and Hall Limited; 1970. pp. 179-180

[31] Arthur J, Napier D. Fifth Symposium on Combustion. London: Chapman and Hall Limited; 1955. p. 303

[32] Gaydon A, Fairbairn A. Fifth Symposium on Combustion. London: Chapman and Hall limited; 1955. p. 24

[33] Greene E, Taylor R, Patterson W. Mechanism of the pyrolysis of acetylene. Journal of Physical Chemistry; 1958;**62**:238

[34] Dubey P, Muthukumarah D, Dash S, Mukhopadhyay R, Sarkar S. Synthesis and characterization of water soluble carbon nanotubes from mustard soot. PRAMANA Journal of Physics. 2005;**65**:681-697

[35] Baska R, Devashankar S, Sarkar R. Analysis of nano carbon obtained from mustard oil soot. International Journal of Emerging Technologies in Computational and Applied Sciences. 2017;**17**:1-5

[36] Novopashin S, Serebrjakova M, Zaikovoki A. Morphology, chemical composition and magnetization of are discharge Fe–C soot. Journal of Engineering and Applied Sciences. 2016;**11**:9130-9133

[37] Kumar D, Paul E, Kavita K, Singh K, Bhatti H. Synthesis and characterization of water soluble carbon nanotubes. Journal of Nanoscience. 2016;**2**:64-65

[38] Klabunde K. Nanoscale Materials in Chemistry. New York: Wiley Interscience; 2001;**1035**:223-262

[39] Krishnamoorti R, Vaia R. Polymer nanocomposites: Introduction in polymer nanocomposites synthesis.

Characterization and hoarding. ACS Symposium. 2002;**804**:1-5

[40] Bandyopadhyay S, Hsieh A, Giannells E. PMMA nanocomposites synthesized by emulsion polymerization. ACS Symposium. 2002;**804**:15-25

[41] Spitaloky Z, Tasis D, Papageli K, Galiotis C. Carbon nanotube polymer composites: Chemistry, processing mechanical and electrical properties. Progress in Polymer Science. 2010;**35**:357-401

[42] Ward H. In: Thomas TR, editor. Rough Surfaces. London: Longman; 1982

[43] Kahnert M, Nousiainen T, Lindquist H, Ebert M. Optical properties of light absorbing carbon aggregates mixed with sulfate assessment of different model geometries for climate forcing calculations. Optical Express. 2012;**20**:10042-10058

[44] Bond T, Bergstrom R. Light absorption by carbonaceous particles: All investigated review. Aerosol Science and Technology. 2006;**40**:27-67

[45] Sara D, Taylor M, Andrew T, Lindsa R, Daniel A, Paola M, et al. Measurement and modeling of the multiwave length optical properties of uncoated flame generated soot. Atmospheric Chemistry and Physics. 2018;**18**:12141-12159

[46] Zhang R, Hu Y, Xu J, Fan W, Chen Z. Flame ability and thermal stability studies of styrene-butmer/graphite oxide nano composite. Polymer Degradation and Stability. 2004;**85**:583-588

[47] Acharya A, Sarwan B, Sharma R, Moghe S, Shrivastav S, Ganesan V. UV-shielding efficiency of TiO_2-polystrene thin films prepared by solution cast method. Frontiers of Physics and Plasma Science-Journal of Physics Conference Series. 2017;**836**:1-5. DOI: 10.10.1088/1742-6596/836/1/012048

[48] Mita A, Isono K. Effective complex refractive index of atmospheric aerosols containing absorbing substances. Journal of the Meteorological Society of Japan. 1980;**58**:69-78

[49] Sabbar A, Mohammed H, Ibrahim A, Saud H. Thermal and optical properties of polystyrene nanocomposites reinforced with soot. Orientation Journal of Chemistry. 2019;**35**:455-460

[50] Russo C, Stanzione F, Alfe M, Cicyolo A, Tregrossi A. Spectral Analysis in the UV-Visible Range for Revealing the Molecular Forms of Combustion Generated Carbonaceous Species. Italy: Chia Laguna; 2011

[51] Gandhi K, Salovey R. Dynamic mechanical behavior of polymers containing carbon black. Polymer Engineering and Science. 1988;**28**:877-887

[52] Ibarra L, Macais A, Palma E. Viscoelastic properties of short of short carbon fiber thermoplastic (SBS) elastomer composites. Journal of Applied Polymer Science. 1995;**57**:831

[53] Nigrawal A, Chand N. Studies on dynamic mechanical analysis and morphology of carbon soot filled in saturated polyester graded composites. International Journal of Science Engineering and Advance Technology. 2014;**2**:116-119

Chapter 7

The Use of Synchronous Fluorescence Technique in Environmental Investigations of Polycyclic Aromatic Hydrocarbons in Airborne Particulate Matter from an Industrial Region in Poland

Aniela Matuszewska and Maria Czaja

Abstract

The applicability of the fluorescence techniques to identify the polycyclic aromatic hydrocarbons (PAHs) in environmental samples is presented. The technique of synchronous fluorescence enabled the identification of the PAHs series containing 2–6 condensed rings in urban airborne particulate matter from Upper Silesia industrial region in Poland. The results obtained by synchronous and conventional fluorescence measurements have been confirmed by those from gas chromatography-mass spectrometry. As the air sample was taken in summer season, the main source of pollution by PAHs component seems to be transport – the exhaust gases from motor vehicles.

Keywords: synchronous fluorescence, polycyclic aromatic hydrocarbons (PAHs), atmospheric pollutions, gas chromatography-mass spectrometry

1. Introduction

Polycyclic aromatic hydrocarbons (PAHs), ubiquitous today in human environment, being harmful as potentially carcinogenic or mutagenic compounds, derive from various sources: natural and anthropogenic ones. Aromatic compounds generated in various natural processes can appear in the natural environment, joining the global cycle of dispersed and concentrated forms of organic matter. The dispersed forms of organic matter are, e.g., hydrocarbons existing in a small amount in rocks and minerals of various origins as well as in abyssal waters getting out on the Earth surface as natural geysers, or hydrocarbons emerging from mining shafts as well as hydrocarbons from volcanic exhalations. The hydrocarbons should be also mentioned from dispersion aureoles round about petroleum and gas deposits. There are also hydrocarbons in under waters accompanying the petroleum, coal, and ore

 IntechOpen

deposits and hydrocarbons penetrating to water reservoirs and to soil as a result of outcrop and erosion of sedimentary rocks. The concentrated forms of the hydrocarbon existence are crude oils, combustible gases, solid bitumens, coals, bituminous shales, and organic-mineral associations of various natures [1]. Among different hydrocarbon groups, there are frequently various PAHs as products of diagenetic, biological, or thermal processes [2–6]. All of these kinds of hydrocarbons occurrence form a geochemical background overlapped by the anthropogenic pollutions [1]. The main sources of anthropogenic PAHs are incomplete combustion of organic matter especially organic fuels (industry, transport, incinerating plants, domestic processes) and thermolysis of organic fossils (coke and asphalt production) [7–9]. All waste organic matter emitters are especially active in an industrial region where they spread to large areas due to various meteorological phenomena. It is because the atmosphere is a zone particularly exposed to pollution.

The permanent contact of living organisms with air makes the polluted atmosphere an especially important area of monitoring. Harmful PAH compounds that pollute the air are very susceptible to adsorption on the other else air pollutant – suspended dust, formed by mineral matter and/or by soot appeared in the air due to lack of or inappropriate protective filters on devices emitting it [10, 11].

The necessity of monitoring of PAHs in air explains the proof of the use much more effective analytical methods enabling their identification in a manner as simple and fast as possible. Among the most frequently applied modern methods of analysis of aromatic hydrocarbons in environmental samples are the chromatographic ones, particularly capillary gas chromatography, the most effective with mass spectrometry as the tool of detection (GC and GC-MS methods) [12–19].

Also high performed liquid chromatographs (HPLC method) are widely applied in these investigations (e.g., [20–22]). However, it is often not cost effective to apply these methods routinely to large numbers of samples.

In the choice of analytical method, it is always very important to take into account the specific properties of analyzed compounds. In the analyzed case, the PAH's high fluorescence efficiency and sensibility as well as simplicity of sample preparation are the reasons that fluorescent techniques are beneficial for PAHs

Figure 1.
Localization of sampling in Upper Silesia Region.

monitoring. However, the conventional fluorescent spectra of multi-component mixture of PAHs have been sometimes difficult to interpret. Now, synchronous technique of fluorescence spectroscopy (SFS) removes these difficulties (e.g., [1, 23]).

The purpose of this work is a presentation of the applicability of the technique of synchronous luminescence for qualitative analysis of aromatic fraction composition of organic substance polluting the urban air in an industrial region. The task sample was taken from Upper Silesia Metropole. This is the region of Poland with many branches of industry, such as mining, metallurgy, electric power station (coke and coal), a dense road network, and a great number of coal-fired home stoves, used sometimes also in the summer. City Mysłowice – the area where air dust was taken to investigations lies centrally in this region (**Figure 1**). The monitoring of the air state in Upper Silesia Metropole and particular cities gives not only an actual information but can also indicate the direction of changes. We believe that it also encourages the people responsible for environment for more intensive efforts to improve air quality.

2. Some remarks on PAHs fluorescence and synchronous fluorescence spectroscopy (SFS)

Fluorescence is the emission of light of wavelength generally different from that of the incident radiation. Most polycyclic aromatic hydrocarbons are fluorescent. It is caused by the fact that the delocalized electrons in the aromatic rings may be easily excited, and the stiff structure does not allow for efficient vibrational relaxation. Fluorescence spectra of each PAH are very characteristic, and they depend on the number and position of aromatic rings. With aromatic ring number increasing, the fluorescence spectrum and emission peak wavelength are all red-shifted from ultraviolet to visible range; the fluorescence emission spectra from one to four rings could be discriminated in the following wavelengths, 275–320 nm, 320–375 nm, 375–425 nm, and 425–556 nm, respectively. It can be used for distinguishing the type of the polycyclic aromatic hydrocarbons (PAHs) as it exists in single type.

The electronic states in the molecules could be singlets or triplets. Singlet state is defined when all the electron spins are paired in the molecular electronic state. The ground state is always singlet (S_0). The excited states could be singlets (S_1, S_2, ...) or triplets (T_1, T_2, ...) when total spin is equal 1. The most probably absorption (A) and emission transitions are singlet-singlet; their intensity is high and luminescence decay time short, 10^{-9}–10^{-7} s. Such emission transition is called fluorescence (F). Molecules in the S_1 state can undergo a spin conversion to the first triplet state T_1; energy of this triplet state is lower than of S_1 state. The emission from T_1, termed phosphorescence (P), is shifted to the longer wavelengths relatively to the fluorescence and characterized by distinctly smaller intensity and longer decay time, 10^{-6}–1 s. It was showed on the simplified Jablonski diagram (**Figure 2**) [24].

The idea of synchronous fluorescence was first suggested by Lloyd and Evett [23]. In conventional fluorescence spectrometry, an emission spectrum is measured for constant excitation wavelength (λ_{exc}). Contrary, the excitation spectrum is measured for constant emission wavelength (λ_{em}), usually to determine the energy levels (or wavelengths) that are responsible for measured emission.
In some sense, excitation spectra replace the absorption spectra. In Constant-Wavelength Synchronous Fluorescence Spectroscopy, the wavelength interval $\Delta\lambda = \lambda_{exc} - \lambda_{em}$ is keeping a constant. For each PAH, the most characteristic λ_{exc} and λ_{em} and in consequence $\Delta\lambda$ values could be choose. Value of $\Delta\lambda$ parameter is so calculated as a difference between effective emission and excitation wavelength characteristic for identified compound. When the difference between the longest

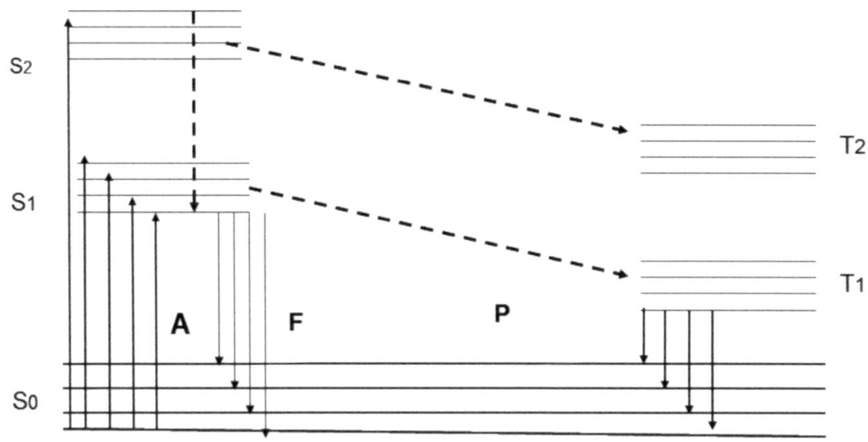

Figure 2.
The simplified Jablonski diagram.

wavelength excitation band and the shortest wavelength emission band is applied as $\Delta\lambda$, there is only one peak in the synchronous fluorescence spectrum [1, 25] The single peak is present at the same wavelength as the longest wavelength excitation band for a synchronous excitation spectrum or the shortest wavelength emission band for a synchronous emission spectrum [26]. The characteristic sets of λ_{exc} and λ_{em} of analyzed compounds could be known from references or obtained independently from measurements of the high purity standards. In this way, synchronous fluorescence spectra are much simpler and easier to analyze than conventional emission. Thus, synchronous fluorescence spectroscopy becomes an attractive alternative for the simultaneous determination of multiple compounds in complex samples. The analytical significance of SFS is confirmed by the fact that this technique is already relatively broadly utilized in investigations of samples of various origins [26–36].

3. Experimental

3.1 The origin and analysis of the investigation object

The analyzed airborne particulate matter derives from Mysłowice urban air (Upper Silesia industrial region, Poland; **Figure 1**). Located in an urban area, a "staplex" high volume sampler with glass fiber filter was applied as the equipment for collecting the airborne particulate matter. The collection of the sample was done in summer (non-heating season). The organic fraction was isolated from the investigated particulate matter by extraction with redistilled n-hexane for 3 hours, using a Soxhlet extractor. The choice of the solvent was substantiated by a mean to obtain nonpolar fraction because many of polar compounds fluoresce. To avoid the eventual fluorescence by traces of these compounds, the additive fractionation was made to reduce their content. For this purpose, Merck's TLC pre-coated plate was used, covered by silica gel layer (thickness of 0.2 mm). The mobile phase for thin layer chromatography process was n-hexane. This procedure enabled to obtain aromatic, aliphatic, and polar fractions. The aromatic fraction was recovered from silica gel by elution with n-hexane using the glass column. For the fluorescence analysis, the solutions of aromatic fraction were prepared with the concentrations in the range from 0.01 to 0.002 mg ml^{-1}.

The same aromatic fraction was also used for further analysis by GC-MS method.

3.2 The apparatus used and experimental conditions

Fluorescence spectra were determined at room temperature using a Jobin-Yvon (SPEX) FLUORLOG 3-12 spectrofluorimeter with a 450 W xenon lamp, a double-grating monochromator, and a Hamamatsu 928 photomultiplier. During measurements, scanning parameters were always put as 0.5 nm per 0.5 s, while the remaining parameters were for each type of measurements as follows: for synchronous measurements, excitation and emission slits have been set to 1 nm; for emission measurements, excitation slit was put to 3 nm, and emission slit – to 1 nm, while for excitation measurements, the slits settings were reversed. To verify the presence of a given PAH, the $\Delta\lambda$ parameter was chosen, according to the reference data [1, 23, 37] as well as to own earlier experimental results.

The comparative analysis of PAHs by gas chromatography-mass spectrometry (GC-MS) was performed using a HP 5890 II gas chromatograph equipped with a fused silica capillary column HP-5 (60 m length × 0.25 mm internal diameter). Helium was the carrier gas used. The GC oven was programmed from 35 to 300°C at a rate of 3°C/min. The gas chromatograph was coupled with a HP 5971A mass selective detector (MSD). The MS was operated with an ion source temperature of 200°C, an ionization energy of 70 eV, and a cycle time of 1 s in the mass range 40–600 Daltons.

4. Results

The results of the qualitative investigations performed are summarized in **Table 1**. Values λ_{ex}, λ_{em}, and $\Delta\lambda$ were collated there for individual PAHs after the experimental and literature data [1] (the limits of differences between the λ data are 1–3 nm). The presence of compounds identified by SFS technique has been confirmed by GC-MS method characterized here by the retention time. The results obtained indicate that both methods have own high research capability, but generally fluorescence analytical procedure seems to be simpler and analytical instrument cheaper in exploitation.

The results of qualitative analysis given in **Table 1** were obtained as a result of record of many synchronous spectra. Several of them are presented below.

4.1 SFS examples of application to the identification of individual aromatics in the analyzed PAH mixture

In **Figure 3**, the synchronous spectrum of analyzed PAH mixture is shown, recorded at the value of $\Delta\lambda$ = 23 nm. This value was proposed by Mille et al. [38, 39] for general characteristics of aromatics from fossil fuels, as an effective parameter to take a fair middle course between the sensitivity and resolving parameters. They used this value for estimation of PAH condensation degree range of investigated aromatic compounds in the mixture. **Figure 3** shows the distribution of emission bands of studied sample in which compounds with a number of rings from 2 to 6 can be identified, thanks to their representation by characteristic bands. The most apparent are substances with 3–5 rings. The bands in the spectrum under discussion identify the following polycyclic aromatic hydrocarbons and its alkyl, mostly methyl derivatives: acenaphthene, benzo(c)fluorene, naphthalenes

Identified compounds	Experimental data $\Delta\lambda$ [nm]	$\lambda_{em}/\lambda_{ex}$ [nm/nm]	Reference data [1] $\Delta\lambda$ [nm]	$\lambda_{em}/\lambda_{ex}$ [nm/nm]	GC/MS retention time [min]
2,3-Dimethyl-naphthalene	54	334/280	54	333/279	14.78
Phenanthrene	53	347/294	53	346/293	25.26
	71	365/294	71	364/293	
	105	357/252	102	356/254	
9-Methyl-phenanthrene	50	349/299	52	350/298	29.11
9,10-Dimethyl-phenanthrene	55	356/301	55	354/299	32.42
	71	355/284	69	354/285	
1-Methyl-7-isopropyl-phenanthrene	67	368/301	67	369/302	38.63
	98	358/260	100	360/260	
1-Methyl-anthracene	20	400/380	21	403/382	28.94
	37	384/347	38	383/345	
9-Methyl-anthracene	22	411/389	23	411/388	29.86
9,10-Dimethyl-anthracene	23	400/377	24	402/378	33.23
	23	424/401	24	424/400	
Fluoranthene	94	418/324	92	417/325	33.96
Benzo(b)fluorene	28	339/311	29	341/312	37.60
	46	356/310	45	357/312	
Pyrene	39	369/330	39	372/333	35.53
	46	378/332	46	379/333	
1-Methyl-pyrene	31	370/339	34	373/339	39.08
4-Methyl-pyrene	52	376/324	51	374/323	39.73
3-Methyl-pyrene	30	378/348	31	376/345	39.94
	50	377/327	48	376/328	
Benzo(c)-phenanthrene	94	376/282	93	374/281	43.37
2-Methyl-benzo-(c)-phenanthrene	66	398/332	66	399/333	43.70
	103	377/274	104	378/274	
Benz(a)-anthracene	23	384/361	23	384/361	44.88
1-Methyl-benz(a)-anthracene	47	389/342	47	389/342	48.45
2-Methyl-benz(a)-anthracene	20	385/365	19	387/368	47.91
7-Methyl-benz(a)-anthracene	17	392/375	18	392/374	48.19
8-Methyl-benz(a)-anthracene	21	387/366	21	388/367	48.77
12-Methyl-benz(a)-anthracene	35	391/356	37	393/356	47.15
1,12-Dimethyl-benz(a)-anthracene	41	404/363	41	402/361	49.35
7,12-Dimethyl-benz(a)-anthracene	53	399/346	52	399/347	51.44
Chrysene	39	362/323	39	361/322	45.16
Benzo(b)-fluoranthene	30	399/369	29	396/367	52.69
	66	430/364	63	430/367	
	94	398/304	95	396/301	

Identified compounds	Experimental data $\Delta\lambda$ [nm] $\lambda_{em}/\lambda_{ex}$ [nm/nm]		Reference data [1] $\Delta\lambda$ [nm] $\lambda_{em}/\lambda_{ex}$ [nm/nm]		GC/MS retention time [min]
Benzo(e)-pyrene	71	390/319	71	389/318	54.37
Benzo(a)-pyrene	34	400/366	33	403/370	54.69
	39	425/386	39	427/388	
Anthanthrene	25	433/408	24	434/410	61.02
	30	433/403	30	434/404	
Benzo(g,h,i)-perylene	50	420/370	50	420/370	62.85
3,4-9,10-Di-benzopyrene	34	433/399	34	432/398	69.21

**The term "luminescence" is used sometimes interchangeably with "fluorescence," describing more widely the discussed phenomenon.*

Table 1.
*Identification by synchronous fluorescence and GC/MS analyses of individual compounds occurring in investigated aromatic fraction of organic mixture desorbed from urban air dust collected in industrial area. Characteristic parameters of individual compounds present in studied aromatic fraction identified on the basis of luminescence * analysis ($\Delta\lambda$, $\lambda_{em}/\lambda_{exc}$) and GC-MS measurements (retention time).*

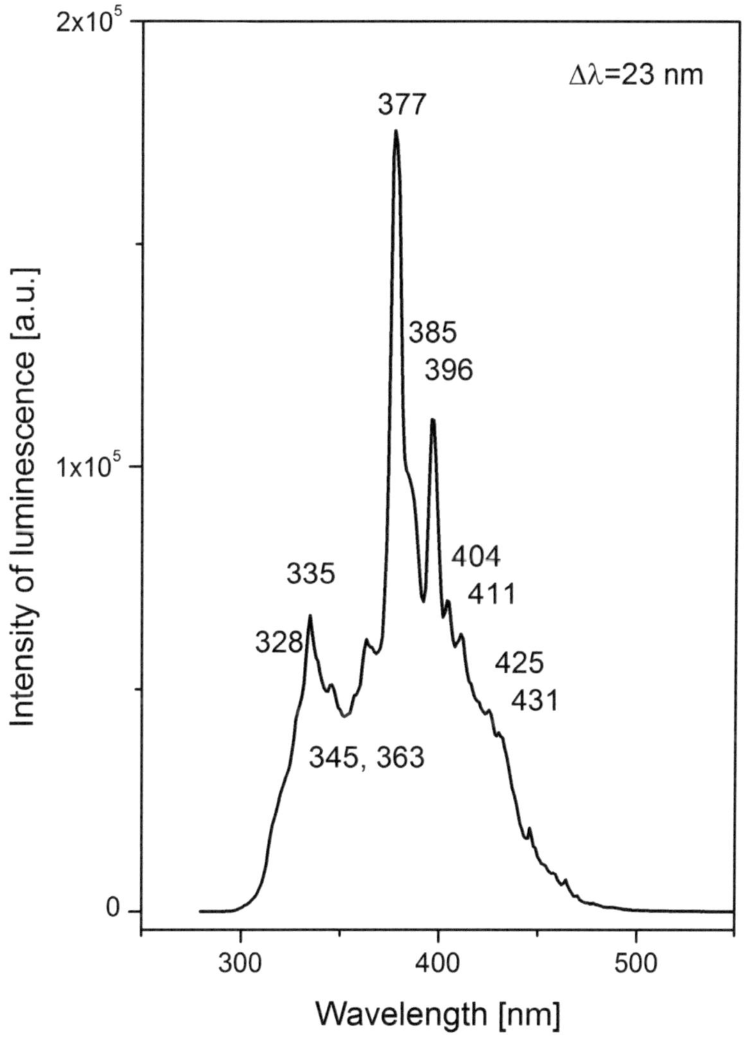

Figure 3.
Synchronous fluorescence spectrum of studied sample measured at $\Delta\lambda$ = 23 nm.

(328 and 335 nm), phenanthrene (345 nm), chrysenes (363 nm), pyrenes, anthracene (377 nm), benzo(a)anthracene (385, 411 nm), benzo(b)fluoranthene (396 and 425 nm), benzo(a)pyrene (404 nm), benzo(g,h,i)perylene (425 nm), and 3,4-9,10-dibenzopyrene (431 nm). In this way, the spectrum recorded at Δλ = 23 nm is a kind of overview spectrum for the composition of the analyzed mixture.

The use of Δλ value calculated for individual compound as it was explained earlier enables the record of a synchronous spectrum with intense band identifying analyzed compound. In **Figure 4,** the spectra are presented and recorded at Δλ = 30 nm, Δλ = 46 nm, and Δλ = 39 nm. Using Δλ = 30 nm, the presence of 3-methylpyrene and benzo(b)fluoranthene was proven on the basis of the most intense bands at 378 and 399 nm, respectively. Moreover, a distinct band of anthanthrene (433 nm) can be seen. In turn, on the spectrum recorded at Δλ = 46 nm, the presence of benzo(b)fluorene (356 nm) is evident, and other distinct bands at 378 and 399 nm correspond to pyrene and 3-methylpyrene, respectively. For chrysene identification, Δλ value equal to 39 nm was taken. In the spectrum obtained, a band at 362 nm characteristic for chrysene is very intense. The band of lower intensity at 369 nm can derive from pyrene. **Table 1** indicates parameter Δλ = 39 nm as characteristic also for benzo(a)pyrene, and in **Figure 3**, the band is seen at 425 nm

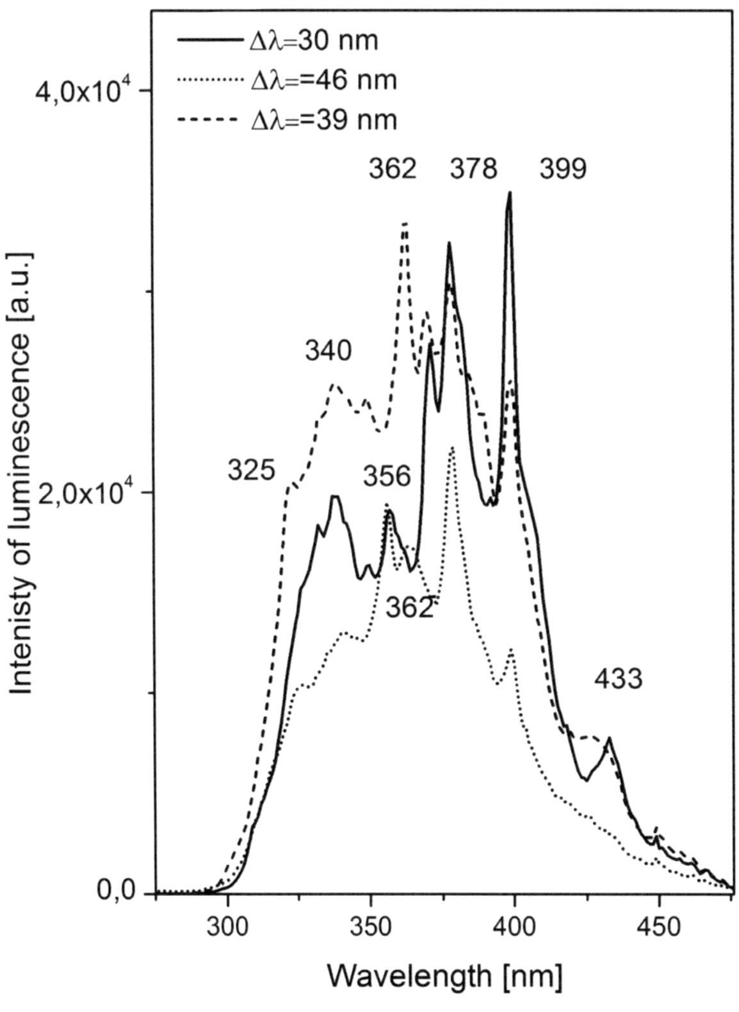

Figure 4.
Synchronous fluorescence spectra of studied sample measured at Δλ = 30 nm, Δλ = 39 nm, and Δλ = 46 nm.

deriving probably from benzo(a)pyrene. The low intensity of benzo(a)pyrene band is undoubtedly a result of low concentration of this compound (see also relatively weak benzo(a)pyrene peak on the GC-MS chromatogram presented below; **Figure 8**). However, in a complex mixture the partial quenching of the fluorescence of individual components is possible by some of other components of the mixture. This process, however, is not yet fully investigated.

Subsequent spectra in **Figure 5** show the presence of phenanthrene, thanks to the characteristic emission bands at 347 nm (at $\Delta\lambda$ = 53 nm) and 357 nm (at $\Delta\lambda$ = 105 nm). Other bands at 376 nm and 399 nm are attributed to 4-methylpyrene and 7,12-dimethylbenzo(a)anthracene.

Figure 6 is an example of verification of the results obtained by synchronous technique, using other spectral fractionation realized by conventional technique of fluorescence. For example, an emission spectrum of chrysene has been recorded at an excitation wavelength of 324 nm (close to the literature value of 322 nm; **Table 1**). Three bands at 361, 370, and 379 nm appeared distinctively complying with the literature data for chrysene: 361, 373, and 381 nm (**Table 1**). The emission spectrum of chrysene is, however, overlapped partly with group of bands at

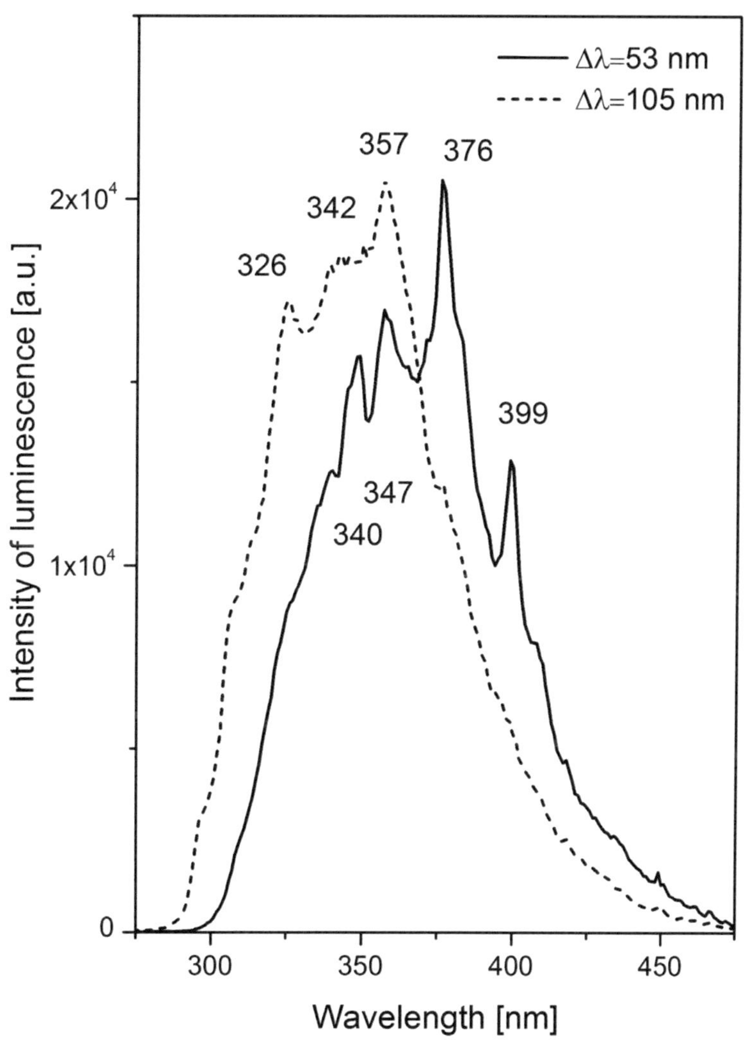

Figure 5.
Synchronous fluorescence spectra of studied sample measured at $\Delta\lambda$ = 53 nm and $\Delta\lambda$ = 105 nm.

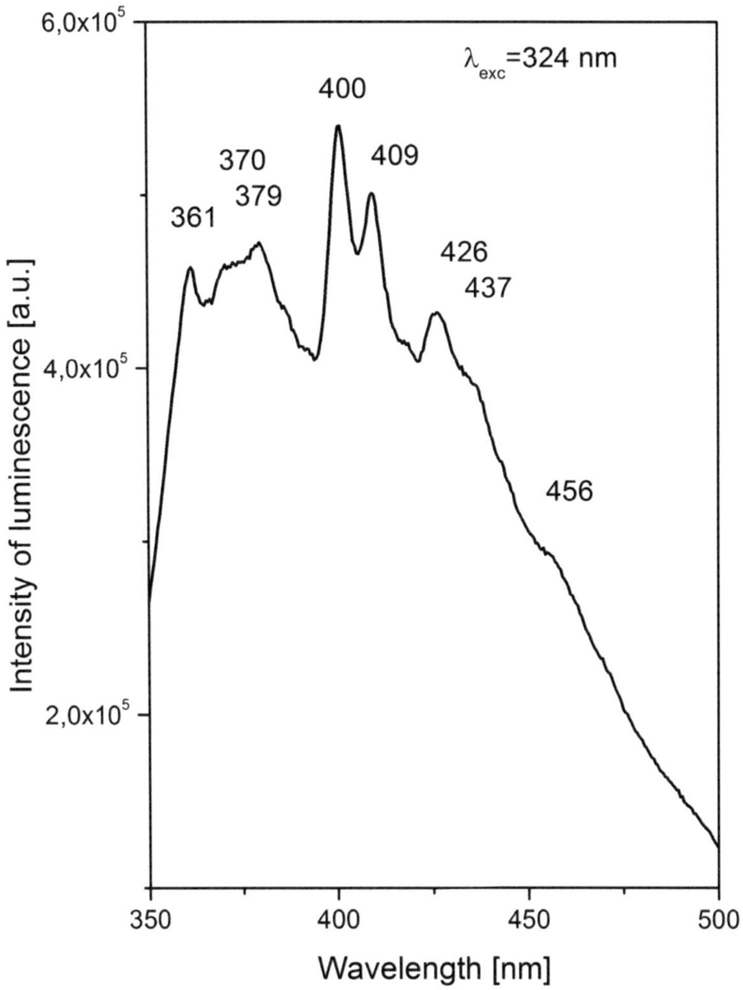

Figure 6.
Fluorescence emission spectrum of studied sample measured at λ_{exc} = 324 nm.

wavelengths: 400, 409, 426, 437, and 456 nm. The origin of these bands seems to have a complex nature. Analysis of the literature data indicates that there are probably several groups of compounds in investigated mixture characterized by similar energy levels corresponding with these emission wavelengths.

Several homologous compounds from the group of anthracene, benzo(a) anthracene, and benzo(a)pyrene can contribute to the mentioned bands. A series of tests with the record of various emission and excitation spectra using conventional techniques and also synchronous one indicated possible contribution to the bands under discussion of 9,10-dimethylanthracene (402, 407, 424, 428, and 453 nm), 1,12-dimethylbenzo(a)anthracene (402, 427, and 455 nm), and benzo(a)pyrene (403, 408, 427, and 431 nm) [1]. However, according to various premises, the benzo(a)pyrene can be the main component of the mentioned group of bands.

To validate this hypothesis, the emission spectrum of benzo(a)pyrene was recorded, using its characteristic λ_{ex} = 350 nm. **Figure** 7 obtained in this manner presents emission bands (402, 409, and 426 nm) of benzo(a)pyrene, confirming contribution of this ingredient to intensity of bands under investigation. An inverse spectral process was also carried out: the spectra recorded at the positions of emission monochromator of λ_{em} = 402 nm and λ_{em} = 427 nm allowed to get characteristic

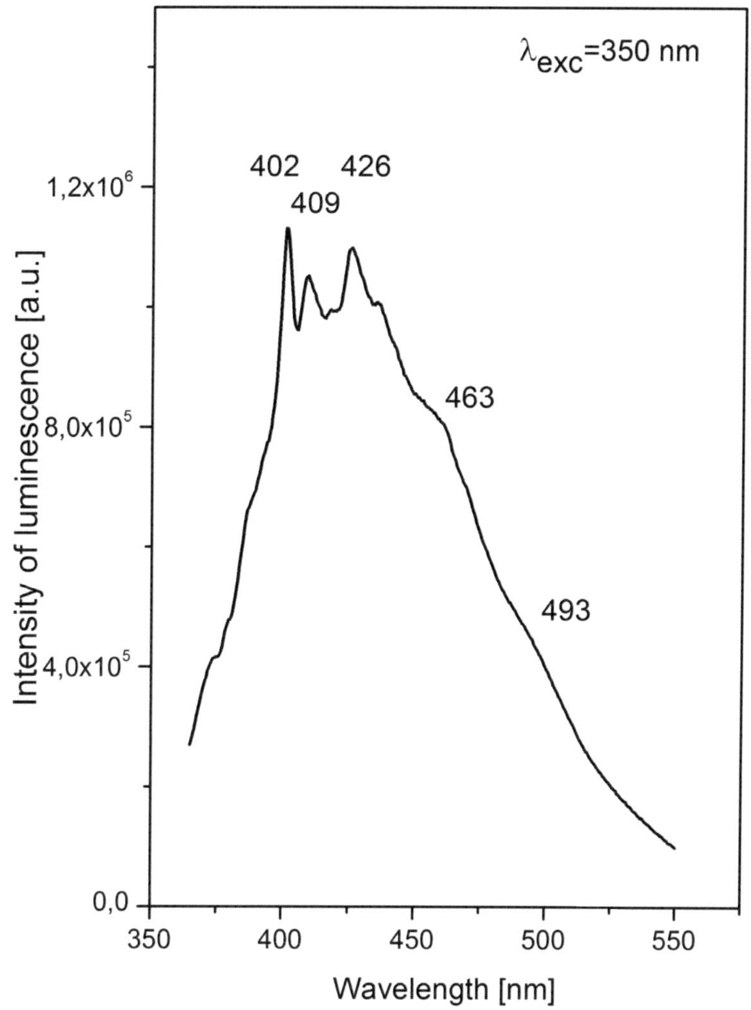

Figure 7.
Fluorescence emission spectrum of studied sample measured at λ_{exc} = 350 nm.

clearly visible excitation bands with maxima at 349 and 372 nm, close to the literature data for λ_{ex} of benzo(a)pyrene at 350 and 370 nm [1]. It is important to mention that in the emission spectrum obtained at λ_{ex} = 350 nm (**Figure 7**), there are also two weak bands at 463 and 493 nm deriving probably from indeno (1,2,3-cd)pyrene.

The high condensed aromatics as coronene were not identified by SFS technique. Lin et al. [26] also found difficulties with the use of SFS to analyze PAHs containing more than six rings. The authors emphasize at the same time that SFS technique is the most sensitive in detecting three- or four-ring PAHs. For the coronene mentioned (seven condensed rings), not identified by SFS, the additional analysis has been performed, and its strong emission band at 446 nm was obtained in the emission spectrum recorded using λ_{ex} = 300 nm.

4.2 Comparative GC-MS analysis

In **Figure 8**, total ion chromatogram (TIC, GC/MS) is presented of aromatic fraction of the extract from analyzed airborne particulate matter. The following PAHs were identified, and their characteristic peaks indicated there by consecutive numbers from 1/ to 20/:

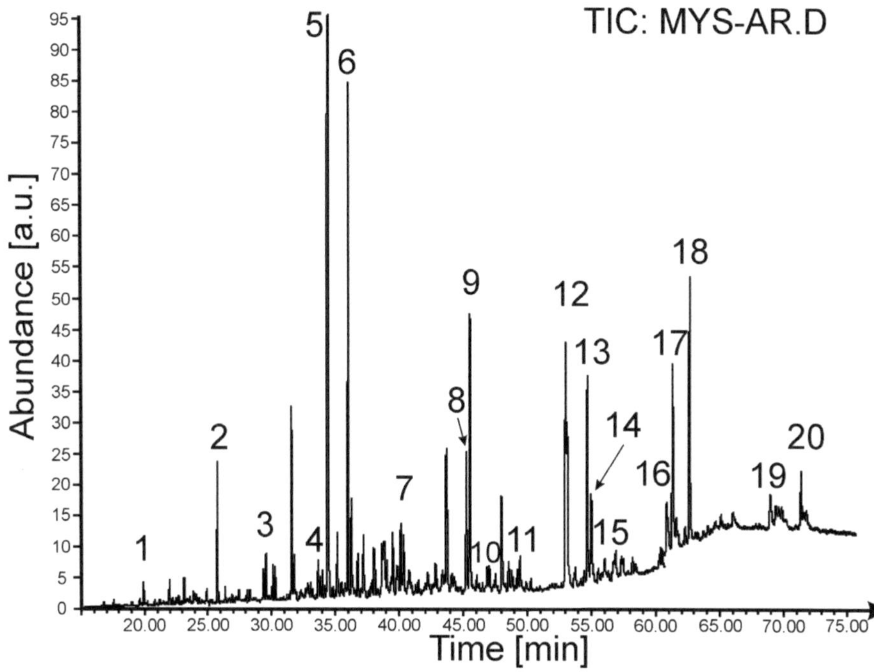

Figure 8.
Total ion-chromatogram (TIC, GC/MS) of aromatic fraction of the extract from analyzed airborne particulate matter.

1/ alkylnaphthalenes, 2/ phenanthrene, 3/ methylphenanthrenes + methylanthracenes, 4/ dimethylphenanthrenes + dimethylanthracenes, 5/fluoranthene, 6/ pyrene, 7/ benzofluorenes + methylpyrenes + benzo(c)phenanthrenes, 8/ benzo(a) anthracene, 9/ chrysene + triphenylene, 10/ alkylbenz(a)anthracenes, 11/alkylchrysenes, 12/benzofluoranthenes, 13/ benzo(e)pyrene, 14/ benzo(a)pyrene, 15/ perylene, 16/ anthanthrene, 17/ indeno(1,2,3-c,d)pyrene, 18/ benzo(ghi)perylene, 19/3,4-9,10-dibenzopyrene, and 20/ coronene. Many compounds from this group seem to be typical for pollution of atmosphere of the investigated industrial region in Poland (e.g., [40]) and at the same time are basic for environmental monitoring in the European Union and the USA (e.g., [41], *loc.cit.*).

The PAHs alkylated are described here as groups of compounds, not all homologs being identified. In **Table 1**, the several alkyl substituted PAHs are presented, among other hydrocarbons, and identified by GC-MS and SFS methods.

5. Discussion of the results

SFS technique made possible the identification of a series of PAHs polluting urban air in the industrial region in Poland. Comparative analysis by GC-MS method has confirmed these results. Qualitative analysis of PAH mixtures by SFS is rather a fast technique and does not need many preparation steps and chemicals. However, for a complex mixture as the analyzed extract from the filter of air dust, the separation into fractions has been done (see subsection 3.1) to reduce the possibilities of mutual quenching of fluorescence by particular components of the mixture. Simplicity of SFS analysis can be shown, e.g., by the possibility of initial estimation of the condensation range of PAHs in analyzed mixture by the use of $\Delta\lambda$ = 23 nm parameter as it was described earlier. However, the most important is here the possibility of receiving the SFS spectra of particular identified PAHs, with one or several well resolved bands, characteristic for analyzed compounds. For recording these spectra, the specific values of Δλ parameters are used after calculation on the basis known spectral characteristics of identified compounds.

The possibility of the use of other analytical techniques of fluorescence method (as the record of emission or extinction conventional spectrum) can broaden the group of identified compounds as, e.g., in the case of coronene with seven condensed rings, not identified by SFS but indicated by emission spectrum as mentioned above.

Thus, the fluorescence techniques, especially SFS, can play a significant role in urban air monitoring for the identification of PAHs. The main order is the environmental impact assessment by the indication of the presence PAHs harmful to health due to toxic, carcinogenic, or mutagenic properties. The total number of PAHs in environment is estimated on about 200, and in exhaust gases, for example, of diesel motors – on over 100 [41]. It is not possible to analyze all of them during everyone monitoring process. It is because only some representative compounds are usually chosen to analyze. For example, Environmental Protection Agency of USA has selected the 16 PAHs for a basic environmental monitoring program [US EPA, Polycyclic Aromatic Hydrocarbons (PAHs)-EPA fact sheet, Washington (DC)-National Center for Environmental Assessment, Office of Research and Development; e.g., [31, 41, 42] *loc. cit.*].

These 16 PAHs are included in the US list of priority pollutants and are usually analyzed: naphthalene, acenaphthene, acenaphthylene, fluorene, phenanthrene, anthracene, fluoranthene, pyrene, benz[a]anthracene, chrysene, benzo[b]fluoranthene, benzo[k]fluoranthene, benzo[a]pyrene, indeno[1,2,3-cd]pyrene, dibenzo[a,h]anthracene, and benzo[ghi]perylene. The influence of these PAHs on the human health has been there also noted. The majority of compounds from this list were identified in the samples of environmental PAHs reported by various authors [1, 12–14, 40, 43]. Their widespread occurrence in the environment causes that most of them are chosen also in European monitoring programs (including Polish programs), e.g., CONCAWE-The Oil Companies European Organization for Environment, Health, and Safety or ARC-International Agency for Research on Cancer (e.g., [41, 44] *loc. cit.*). These compounds characterize very high toxicity.

Almost all of compounds from EPA list are identified among PAHs analyzed here. Taking into account relative intensities of pollutant peaks seen clearly in the ion-chromatogram realized for this work (**Figure 8**), one can stated that the most intense peaks can be attributed to fluoranthene and pyrene, not carcinogenic after EPA. The highest carcinogenicity is attributed, in turn, to benzo(a)pyrene and dibenzo(ah)anthracene, relatively lower to benzo(b)fluoranthene and the lowest to chrysene (potentially carcinogenic) and also to benzo(a)anthracene and indeno(1,2,3-cd)pyrene. Other compounds from the EPA list are not noted as carcinogenic. Main representative of carcinogens – benzo(a)pyrene does not dominate in analyzed group of PAHs, what may be related to summer season of sampling [1], but its content in air of the Silesian industrial region is still relatively high [44].

The general similarity of own results with these ones from the world literature data can suggest that major of identified PAHs may be universal products of high chemical stability, deriving from incomplete burning of various organic fuels. It may be also possible that these compounds derive mainly from the incomplete burning of oil fuels as the result of so-called low emission from vehicle exhaust because of widespread development of automotive transport. In the case of our own results, it could be confirmed by PAH distribution of the summer season when low emission pollutions are dominated by the exhaust gases from car engines as it is suggested below.

Considerable air pollution in the investigated urban atmosphere results in undoubtedly high pollution of whole region of Silesia. This is due to the high degree of industrialism, where an emission of diverse pollutants from many sources can be expected. In the region mentioned, the certain branches of industry, such as mining, metallurgy, nonferrous metals, and coke-making industries were developed excessively. There are also electric power stations and heat generating plants. High intensification of motor transport increases additively level of pollution in this region [13, 14].

The sources of air pollution in the Mysłowice urban area create a specific character of this air compared to that in the whole industrial region. The investigated air dust was sampled in summer. In this season, industrial production using organic fuels is underway, but the low emission from home stoves is minimalized. The main sources of low emission, however, seem to be in summer the engine fumes especially these from diesel motors (despite various actions to improve them). After the literature data [43], in summer season, one can expect in Silesia the relative increasing of concentration of pollutions originating from automobile exhaust as, e.g., pyrene, chrysene, triphenylene, benzo(g,h,i)perylene, benzo(c) phenanthrene, benzo(k)fluoranthene, and indeno(1,2,3-cd)pyrene. All of these compounds have been identified in analyzed PAH mixture, and they show clear dominance over other compounds (**Figure 8**). The domination of the engine exhaust in analyzed air pollution might be caused by the intense automobile traffic on the city area. There are three great car transit lines through the city: a highway autostrada passing by the city, expressway, and national road crossing the city center (the beltway around the city is only enabled in spatial planning).

The particular identified pollutants or their groups can thus help to indicate the pollution sources and to look for ways to improve the quality of environment.

6. Conclusions

The use of synchronous fluorescence spectroscopy (SFS technique) is advantageous for qualitative monitoring of the PAHs mixtures from urban air of the industrial region because this technique is simple, fast, low time consuming, and of low costs. Simpler and better separated spectra may be obtained in this manner than in the case of conventional fluorescence analysis. The special advantage of SFS technique is the possibility of analysis of complex environmental PAHs mixtures without the need of multistage sample fractionation. In this case, the "separation" of the complex mixtures is performed spectroscopically, which may be called the "spectral fractionation technique" [1].

Technique SFS used for samples in a form of solution, applied at the ambient temperature, allows obtaining well-separated spectra, simplified to one or several bands, allowed identification of individual compounds. This analytical process is possible, thanks to the use of the parameter $\Delta\lambda$ characteristic for the particular identified compound. In this manner, the presence was stated in analyzed urban air dust of the PAHs with 2–6 condensed rings. However, the most clearly marked were aromatics from groups of anthracenes, benzo(a)anthracenes, pyrenes, and benzo(a)pyrenes. These compounds are among these PAHs, which characterize high fluorescence yield. The compounds from the group of naphthalene, benzofluorene, phenanthrene, benzophenanthrene, chrysene, fluoranthene, benzofluoranthene, and anthanthrene were also identified. The verification was performed for specific compounds, by recording of respective conventional emission or excitation spectra.

The results obtained may be a basis to discussion on the evaluation of the environmental hazards.

The harmful influence of identified PAHs is a reason why they should be of a special interest in investigations of environmental samples. Thus, the choice of the effective analytical methods and techniques is here very important. The synchronous fluorescence technique seems to be appropriate method for monitoring PAHs in air dust samples of the industrial regions, especially when combustion of organic fossils still remains the main source of energy.

Abbreviations

SFS	synchronous fluorescence spectroscopy
PAHs	polycyclic aromatic hydrocarbons
TIC	total ion-chromatogram
TLC	thin layer chromatography
GC/MS	gas chromatography-mass spectrometry
HPLC	high performed liquid chromatography
MSD	mass selective detector
EPA	Environmental Protection Agency

Author details

Aniela Matuszewska* and Maria Czaja
Faculty of Natural Sciences, Institute of Earth Sciences, University of Silesia, Sosnowiec, Poland

*Address all correspondence to: aniela.matuszewska@us.edu.pl

References

[1] Aleksiejewa TA, Tieplickaja TA. Spektrofluorimetriczeskije metody analiza aromaticzeskich uglewodorodow w prirodnych i technogennych sredach. Leningrad (russ.): Gidrometeoizdat; 1981

[2] Knorr M, Schenk D. About the question of synthesis of polycyclic aromatic hydrocarbons by bacteria. Archiv für Hygiene und Bakteriologie. 1968;**152**:282-285

[3] Bojakowska I, Sokołowska G. Polycyclic aromatic hydrocarbons in crude oils from Poland. Geological Quarterly. 2001;**45**:81-86

[4] Radke M, Willsch H, Teichmüller M. Generation and distribution of aromatic hydrocarbons in coals of low rank. Organic Geochemistry. 1990;**15**:539-563

[5] Santana Rodriguez JJ, Padrón SC. Fluorescence techniques for the determination of polycyclic aromatic hydrocarbons in marine environment: An overview. Analusis. 2000;**28**(8):710-717

[6] Rushdi AI, Simoneit BRT. Hydrothermal alteration of the Northeastern Pacific Ocean. Part 1. Applied Geochemistry. 2002;**17**:1401-1428

[7] Howsam M, Jones KC. Sources of PAHs in the environment. In: Neilson AH, editor. The Handbook of Environmental Chemistry, vol 3.1: PAHs and Related Compounds—Chemistry. Berlin, Heidelberg, New York: Springer; 1998. pp. 137-174

[8] Ravindra K, Sokhi R, Grieken RV. Atmospheric polycyclic aromatic hydrocarbons: Source attribution, emission factors and regulation. Atmospheric Environment. 2008;**42**:2895-2921

[9] Abdel-Shafy HI, Mansour MSM. A review on polycyclic aromatic hydrocarbons, environment, impact, effect on human health and remediation. Egyptian Journal of Petroleum. 2016;**25**:107-123

[10] Desantes JM, Berrmúdez V, Garcia JM, Fuentes E. Effects of current engine strategies on the exhaust aerosol particle size distribution from a heavy-duty diesel engine. Journal of Aerosol Science. 2005;**36**:1251-1276

[11] Maricq MM. Chemical characterization of particulate emission from diesel engines: A review. Journal of Aerosol Science. 2007;**38**:1079-1118

[12] Levy JM, Dolata LA, Ravey RM. Considerations of SFE for GC/MS determination of polynuclear aromatic hydrocarbons in soils and sediments. Journal of Chromatographic Science. 1993;**31**:349-352

[13] Bodzek D, Luks-Betley K, Warzecha L. Research on the determination of toxic organic compounds in airborne particular matter in Upper Silesia. Polish Journal of Environmental Studies. 1993;**2**:13-22

[14] Bodzek D, Luks-Betley K. Warzecha L, Determination of particle-associated polycyclic aromatic hydrocarbons in ambient air samples from the Upper Silesia region of Poland. Atmospheric Environment. 1993;**27A**:759-764

[15] Węglarz A, Skrok R. Application of GC-FID and GC-MS for assessing PAHs in suspended dust. Central European Journal of Public Health. 2000;**8**:86-88

[16] Poster DL, Schantz MM, Sander LC, Wise SA. Analysis of polycyclic aromatic hydrocarbons (PAHs) in environmental samples: A critical review of gas chromatographic (GC) methods.

Analytical and Bioanalytical Chemistry. 2006;**386**:859-881. DOI: 10.1007/s00216-006-0771-0

[17] Krupadam RJ, Bhagat B, Khan MS. Highly sensitive determination of polycyclic aromatic hydrocarbons in ambient air dust by gas chromatography-mass spectrometry after molecularly imprinted polymer extraction. Analytical and Bioanalytical Chemistry. 2010;**397**(7):3097-3106. DOI: 10.1007/500216-010-3858-6

[18] Elorduy I, Durana N, Garcia JA, Gómez MC, Alonso L. Optimization and validation of thermal desorption gas chromatography-mass spectrometry for the determination of polycyclic aromatic hydrocarbons in ambient air. Journal of Analytical Methods in Chemistry. 2018. DOI: 10.1155/2018/8734013

[19] Wang M, Huang R-J, Cao J, Dai W, Zhou J, Lin C, et al. Determination of n-alkanes, polycyclic aromatic hydrocarbons and hopanes in atmospheric aerosol: Evaluation and comparison of thermal desorption, solvent extraction, GC-MS approaches. Atmospheric Measurement Techniques. 2019;**12**:4779-4789

[20] Bacaloni A, Cafaro C, de Giorgi L, Ruocco R, Zoccolillo L, Durana N. Improved analysis of polycyclic aromatic hydrocarbons in atmospheric particulate matter by HPLC-fluorescence. Annali di Chimica. 2004;**94**(9-10):751-759

[21] Veta I, Sekiguchi N, Fujimura K, Yoshimura T, Narukami S, Mochizuki S, et al. Determination of airborne polycyclic aromatic hydrocarbons by HPLC using SPE-type collection device. Chromatography. 2018;**39**:119-124

[22] Assenmacher-Maiworm H, Hahn J-U, Heinrich B, Schuh C, Hebisch R, Brock TH, et al. Policyclic aromatic hydrocarbons (PAHs) – Method for the determination of semi-volatile PAHs in work place air using high performance liquid chromatography. Air Monitoring Methods. 2018. DOI: 10.1002/3527600418.amOpahse 191

[23] Lloyd JBF, Evett IW. Prediction of peak wavelengths and intensities in synchronously excited fluorescence emission spectra. Analytical Chemistry. 1977;**49**:1710-1715

[24] Lakowicz JR. Principles of Fluorescence Spectroscopy. 3rd ed. Boston: Springer; 2006. p. 954

[25] Vo-Dinh T. Synchronous luminescence spectroscopy: Methodology and applicability. Applied Spectroscopy. 1982;**36**:576-581

[26] Lin ELC, Cormier SM, Racine RN. Synchronous fluorimetric measurement of metabolites of polycyclic aromatic hydrocarbons in the bile of brown bulhead. Environmental Toxicology and Chemistry. 1994;**13**:707-715

[27] Abbott DW, Moody RL, Mann RM, Vo-Dinh T. Synchronous luminescence screening for polynuclear aromatic compounds in environmental samples collected at a coal gasification process development unit. American Industrial Hygiene Association Journal. 1986;**47**(7):379-385

[28] Mastral AM, Callén M, Mayoral C, Galbán J. Polycyclic aromatic hydrocarbon emissions from fluidized bed combustion of coal. Fuel. 1995;**74**:1762-1766

[29] Matuszewska A, Czaja M. The use of synchronous luminescence spectroscopy in qualitative analysis of aromatic fraction of hard coal thermolysis products. Talanta. 2000;**52**:457-464

[30] Matuszewska A, Czaja M. Aromatic compounds in molecular phase of Baltic amber – synchronous luminescence analysis. Talanta. 2002;**56**:1049-1059

[31] Santana Rodriguez JJ, Padrón SC. Fluorescence techniques for the determination of polycyclic aromatic hydrocarbons in marine environment: An overview. Analusis. 2000;**28**(8):710-717

[32] Patra D, Mishra A. Total synchronous fluoresence scan spectra of petroleum products. Analytical and Bioanalytical Chemistry. 2002;**373**(4-5):304-309. DOI: 10.1007/SOO216-002-1330-y

[33] Li Y-Q, Li X-Y, Shindi AAF, Zou Z-X, Liu Q, Lin LR. Synchronous fluorescence spectroscopy and its applications in clinical analysis and food safety evaluation. In: Geddes CD, editor. Reviews in Fluorescence. New York: Springer Science; 2010. DOI: 10.1007/978-1-4419-9828-6_5

[34] Zhang R, Yuan Q, Chen K, He L. Constant-wavelength synchronous fluorescence spectrometry for the determination of polycyclic aromatic hydrocarbons in water samples. Advanced Materials Research. 2012;**490-495**:3202-3206

[35] Sharma H, Jain VJ, Khan ZH. Use of constant wavelength synchronous spectrofluorimetry for identification of polycyclic aromatic hydrocarbons in air particulate samples. Spectrochimica Acta Part A. 2013;**108**:268-273. DOI: 10.1016/j.saa.2013.01.079

[36] Nógrega JA, dos Santos RL, de Nara RH, Nunes de Oliveira RA, Capato CF, Falcão EA, et al. Quantification of chrysene and benzo(a)pyrene in surface water samples by fluorescence measurement. Brazilian Journal of Development. 2020;**6**(1):3092-3104

[37] Kershaw JR. Fluorescence spectroscopy in the characterization of coal-derived liquids. Fuel. 1978;**57**:299-303

[38] Mille G, Kister J, Guiliano M, Dou H. Spectroscopie de fluorescence UV: Technique d`excitation émission synchrones. Applications á l`étude de composés fossiles. Spectra. 1985;**13**(106):27-31

[39] Mille G, Guiliano M, Kister J. Analysis of evolution of coals: UV fluorescence spectroscopy study. Organic Geochemistry. 1988;**13**:947-952

[40] Marynowski L, Pięta M, Janeczek J. Compositoon and source of polycyclic aromatic compounds in deposited dust from selected sites around the Upper Silesia. Geological Quarterly. 2004;**48**:169-180

[41] Mazur-Badura X. Profil WWA w cząstkach stałych (PM) emitowanych z silnika Diesla. Nafta-Gaz. 2012;**68**: 246-253. (in Polish)

[42] Fazio T, Howard JW. In: Bjørseth A, editor. Polycyclic Aromatic Hydrocarbons. New York: Marcel Dekker; 1983. pp. 464-468

[43] Warzecha L, Luks-Betlej K, Bodzek D. Metody analizy wielopierścieniowych węglowodorów aromatycznych występujących w naturalnym środowisku człowieka. Chemist-Analyst. 1988;**33**:21-54. (in Polish)

[44] Rusin M, Marchwińska-Wyrwał E. Zagrożenia zdrowotne związane ze środowiskowym narażeniem na wielopierścieniowe węglowodory aromatyczne (WWA). Medycyna Środowiskowa. 2014;**17**(3): 7-13. (in Polish)

Chapter 8

Qualitative Characterisation of Trace Elements in Diesel Particulate Matter from In-Use Diesel Engine Passenger Vehicles by Means of Laser-Induced Breakdown Spectroscopy

Richard Viskup, Christoph Wolf and Werner Baumgartner

Abstract

In this research, we applied laser-plasma spectroscopy technique for the measurement of trace chemical elements in the exhaust emissions generated from in-use diesel engine passenger vehicles. We use high resolution laser-induced breakdown spectroscopy (LIBS) technique for diagnostics of soot and particulate matter (PM). Here we analysed soot and PM, extracted from exhaust manifold part, from different passenger vehicles that are used in daily life environment. The main aim of this study is to reveal the trace chemical elements in different PM matrices. The presence of trace elements in exhaust emissions can originate from different sources: from injected fuel type and fuel additives, engine lubricants, engine combustion process, incomplete catalytic reaction, inefficiency or wear out of PM filtering devices, dysfunctions or failures of engine or vehicle or even information related to polluted intake air.

Keywords: laser-induced breakdown spectroscopy (LIBS), particulate matter, soot, nanoparticles, emissions, emission standards, diesel, diesel engine, diesel vehicles, in-use vehicles, trace elements

1. Introduction

In this research, the laser-induced breakdown spectroscopy (LIBS) technique for diagnostics of trace chemical elements in diesel particulate matter (DPM) formed from in-use diesel engine passenger vehicles has been used [1–3].

Laser-induced breakdown spectroscopy is a powerful spectrochemical measurement technique for fast qualitative and very sensitive quantitative compositional analysis of various forms of matter: solid state, liquid, gas as well as fine powders or nanoparticles [4–6]. One of the pioneers in measurement of particulate trace emissions from vehicles was the group of Schauer et al. [7] as they used a comprehensive dilution source sampler, organic chemical analysis and X-ray fluorescence analysis for mass and chemical composition measurements of fine particles. Other groups [8–11] used inductively coupled plasma mass spectrometry ICP-MS and XRF for characterisation of metals and other particle-phase species from on-road motor vehicles. They found the following trace elements in the particles: Al, Ba, Be, Ca, Cd, Co, Cr, Cu, Fe, K, Mg, Mn, Mo, Na, Ni, Pb, Pt, S, Sr, Ti, V and Zn. Other groups [12–16] used ICP-OES to characterise the different biodiesel samples with special concern to quantify the Al, Ca, Cu, Fe, K, Mg, Mn, Na, Ni, P, Sr, B and Cl content, to evaluate the fuel quality and to control the emission of pollutants to the atmosphere. In this case, the samples were prepared using a high pressure asher digestion procedure for metal determination in biodiesel samples. Different groups used ICP-MS to characterise additional bound elements, such as Cd, As, Ba and Ti in the particulate matter collected from ultra-low-sulphur diesel and biodiesel powered engine exhaust emissions [17].

In the past two decades, a new laser technology evolved in the application of laser-induced breakdown spectroscopy into combustion diagnostics. One of the first research papers that reported the LIBS for on-line engine equivalence ratio measurement was performed by the group of Ferioli et al. [18], followed by research related to measurements of hydrocarbons using LIBS [19]. The implementation of LIBS for in-cylinder measurements was made by Joshi et al. [20], followed by Groß et al. [21]. Another group has studied the LIBS to monitor local lambda values during mixture formation in a direct-injection engine [22]. Different application of LIBS to an engine valve has been used by the group of Lopez-Quintas et al. [23] for mapping of mechanical specimens. Kiefer et al. [24] have used laser-induced breakdown spectroscopy in a partially premixed turbulent jet flame, and the group of Hsu et al. [25] performed measurements of fuel air-ratio in methane-air flames at different pressures. The qualitative and quantitative characterisation of major chemical elements bound in different types of diesel particulate matter measured by laser-induced breakdown spectroscopy technique has been studied by Viskup et al. [26, 27]. And the identification of minor chemical elements in diesel particulate matter by LIBS was studied by Viskup et al. [28].

In this research, the main aim is to measure the trace chemical elements in particulate matter from diesel exhaust emissions. The presence of trace elements in PM can reveal different types of information related to vehicle and engine, combustion process, injected fuel type, fuel additives, engine lubricants, state of selective catalytic reduction devices, inefficiency or wear out of PM filtering devices, engine failure, engine wear out or information related to polluted intake air.

2. Experimental

2.1 Laser-induced breakdown spectroscopy setup

The laser created plasma was generated by the Nd:YAG solid state laser from Quantel corp. This laser operates at the fundamental wavelength of 1064 nm with a pulse duration of 8.5 ns and a laser energy of 300 mJ. The emerged laser radiation has been focused with 10 cm focusing lens into the plane solid target surface to create a plasma. Optical emission from plasma has been collected perpendicularly via optical telescope into the high resolution Echelle spectrograph model Aryelle Butterfly from LTB Berlin equipped with sensitive ICCD detector. The spectrometer consists of two separate spectrographs—UV part and VIS part. The UV part is from 190 to 440 nm, and the VIS part of the optical spectrum is from 440 to 800 nm. Spectral resolution is from 3 pm (picometre) to 7 pm for VUV and from 4 pm to 8 pm for VIS part. This Echelle spectrograph provides spectral information in broad range with very high resolution. Optical emission from plasma has been collected

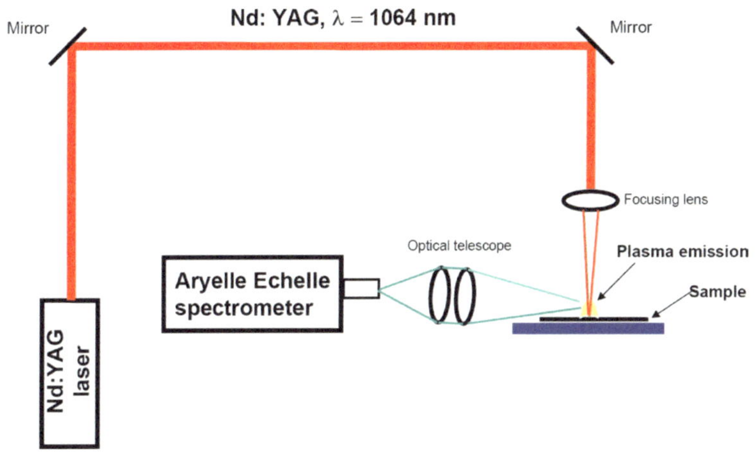

Figure 1.
Layout of laser-induced breakdown spectroscopy setup.

from VUV as well as from VIS parts; thus, the total spectral window from 190 to 800 nm wavelength has been recorded. The delay time of 1 μs after the laser trigger and the gate width of 2 μs were used. The LIBS measurements were performed in open air atmosphere at atmospheric pressure and at room temperature. Layout of the Laser Induced Breakdown Spectroscopy setup is shown in the **Figure 1**.

2.2 Sample preparation and collection

Different PM samples from in-use diesel engine passenger vehicles of major brand car producers in Europe have been analysed by LIBS. Particulate matter has been collected from the tail pipe at the end of the exhaust manifold, while selections of in-use vehicles were performed randomly. Laser-induced breakdown spectroscopy from DPM shows optical emission and spectral lines that are characteristic in ultraviolet and visible spectral region. The collected particulate matter from diesel engine passenger vehicles and exhaust manifold has been mechanically pressed into small pellets with flat disc shape. Each displayed spectrum has been averaged over 12 laser shots.

3. Results and discussion

3.1 Identification of the major chemical elements in PM

Characteristic laser-induced breakdown spectroscopy signal from measurement of diesel particulate matter is shown in **Figure 2**. LIBS spectra generated from particulate matter collected from in-use diesel engine passenger vehicles exhibit characteristic optical emission lines with distinct of atomic, ionic and molecular origin included in the signal. Strong optical emission is from major spectral lines, particularly from carbon, iron, magnesium, aluminium, chromium, zinc, sodium and calcium. These elements were in previous research identified in PM as major components of diesel particulate matter [26].

By means of high-resolution optical emission LIBS spectroscopy, different PM matrices were spectrochemically analysed. From analytical measurements, the composition of major chemical elements in the particulate matter collected from different in-use diesel engine passenger vehicles was obtained. From qualitative

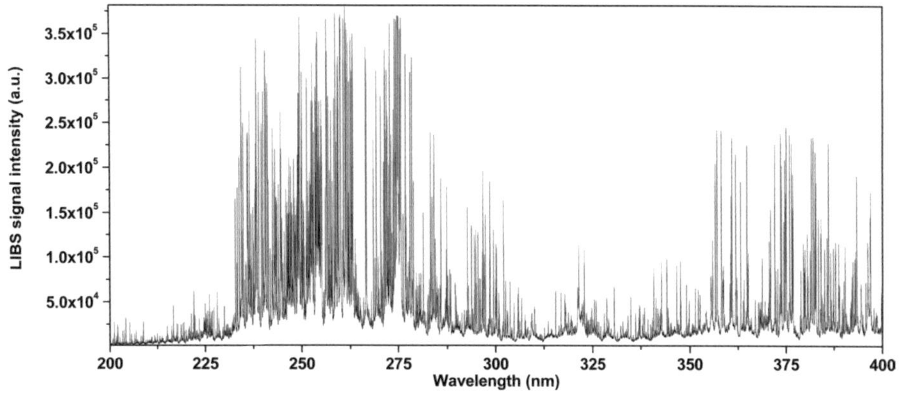

Figure 2.
Optical emission LIBS spectra from diesel particulate matter sample. High intensity spectral lines are from major components carbon, iron, magnesium, aluminium, chromium, zinc, sodium and calcium.

measurements and calibration curves, we found that the major chemical elements of DPM besides the carbon are iron, magnesium, aluminium, chromium, zinc, sodium and calcium with different concentrations. By using quantitative analytical LIBS approach, the maximum concentrations of major chemical elements in DPM from in-use Diesel engine passenger vehicles were measured as follows: Carbon up to ~ 64 weight percent (wt%), Fe ~ 54 wt%, Mg ~ 7 wt%, Al ~ 6 wt%, Cr ~ 6 wt%, Zn ~ 7 wt%, Na ~ 11 wt%, Ca ~13 wt%, for more details see Ref. [26, 27].

3.2 Identification of the minor chemical elements in PM

Further research was dedicated to identify the minor chemical elements of DPM. The state-of-the-art laboratory LIBS setup has been used to obtain high-resolution optical emission spectra images. The qualitative measurements and LIBS signal show the minor chemical elements with optical emission spectra from silicon, nickel, titan, potassium, strontium and molybdenum. More detail study of this topic is presented in Ref. [28].

3.3 Identification of the trace elements in PM

To identify the trace elements in various DPM matrices, the LIBS setup was optimised for optical detection to obtain high-quality spectral data. Acquired signals show optical emissions from trace elements, particularly from barium, boron, cobalt, copper, phosphorus, manganese and platinum in high resolution and in good signal-to-noise ratio. Optical emission spectra from atomic and ionic lines of identified trace elements are shown in **Figure 3**. Here we only select few samples with most pronounced signal to clearly visualise the peak line shape of spectral information.

3.4 Trace elements of diesel particulate matter

Barium spectral line: ionic emission from Ba II @ 455.40 nm is shown in **Figure 3(a)**. In this figure, a raw spectral data from LIBS measurements are shown. Here we select six different diesel particulate matter samples with most intense Barium peak. Selected samples with barium ionic line are samples with numbers 61, 4, 51, 26, 41 and 60. Barium signal has been measured in 26 samples, from 67 different DPM samples, see **Figure 4(a)**. From LIBS spectra, one can observe that analysed signal mainly line peak shape, line peak intensity and line peak width is

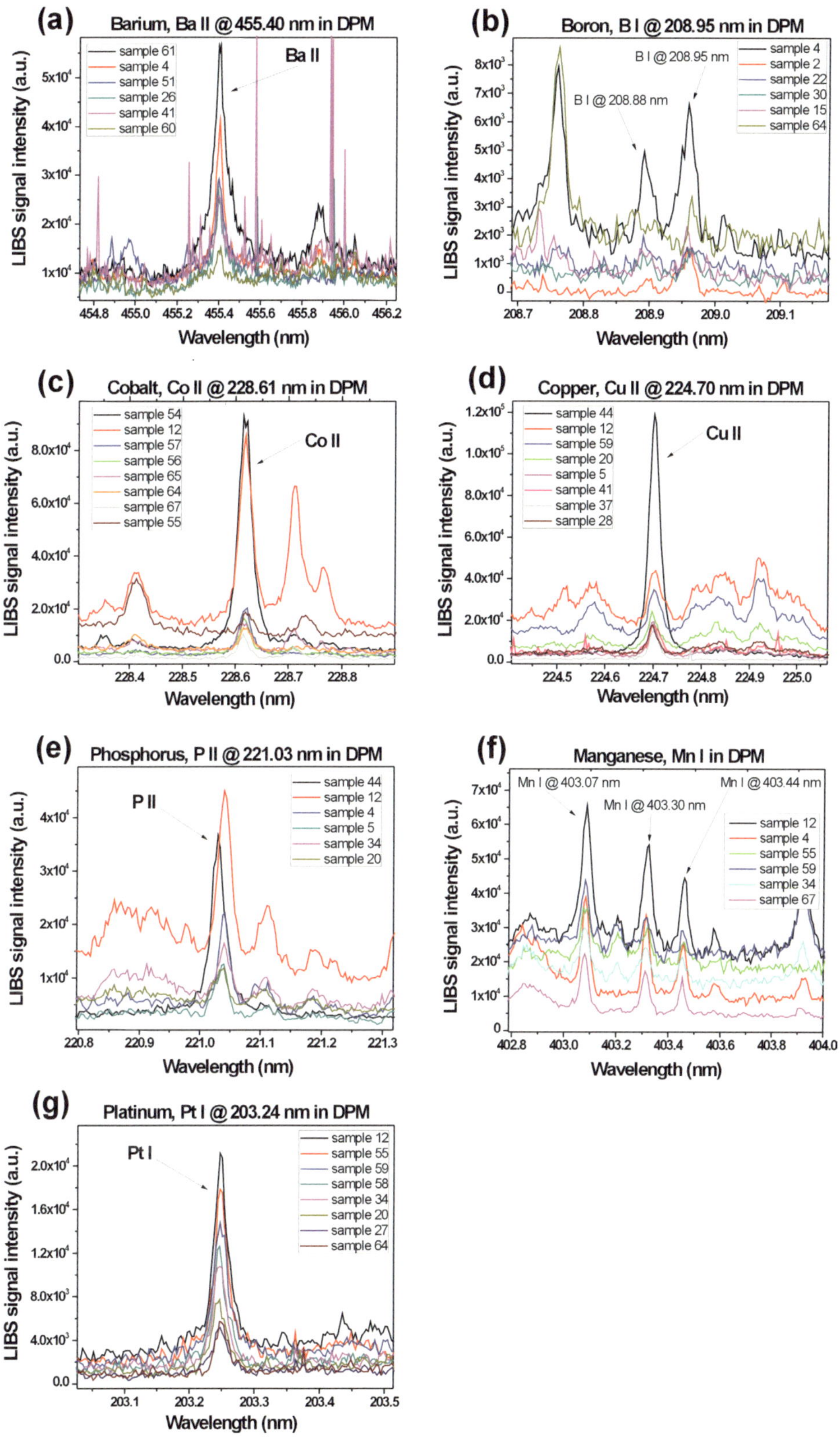

Figure 3.
Optical emission spectra from barium (a), boron (b), cobalt (c), copper (d), phosphorus (e), manganese (f) and platinum (g), measured by high-resolution LIBS technique from diesel particulate matter collected from in-use passenger diesel engine vehicles.

changing according to different DPM samples. The strength of particular atomic or ionic line is basically proportional to the concentration of the element in the studied material. Thus, for qualitative comparison, we numerically calculate the respond signal—the integral of peak values for each spectral line of interest—to obtain information about elemental atomic composition of different types of diesel particulate matter. The results from numerical calculation and integration of peak area are shown in **Figure 4**. From this figure, one can easily compare the individual changes in trace signal related to concentration values in a.u. (arbitrary unit). Nevertheless, in case of exact quantitative characterisation of the trace element in DPM, the particular calibration of trace element signal would be necessary to perform. However, from previous analytical LIBS measurements and qualitative

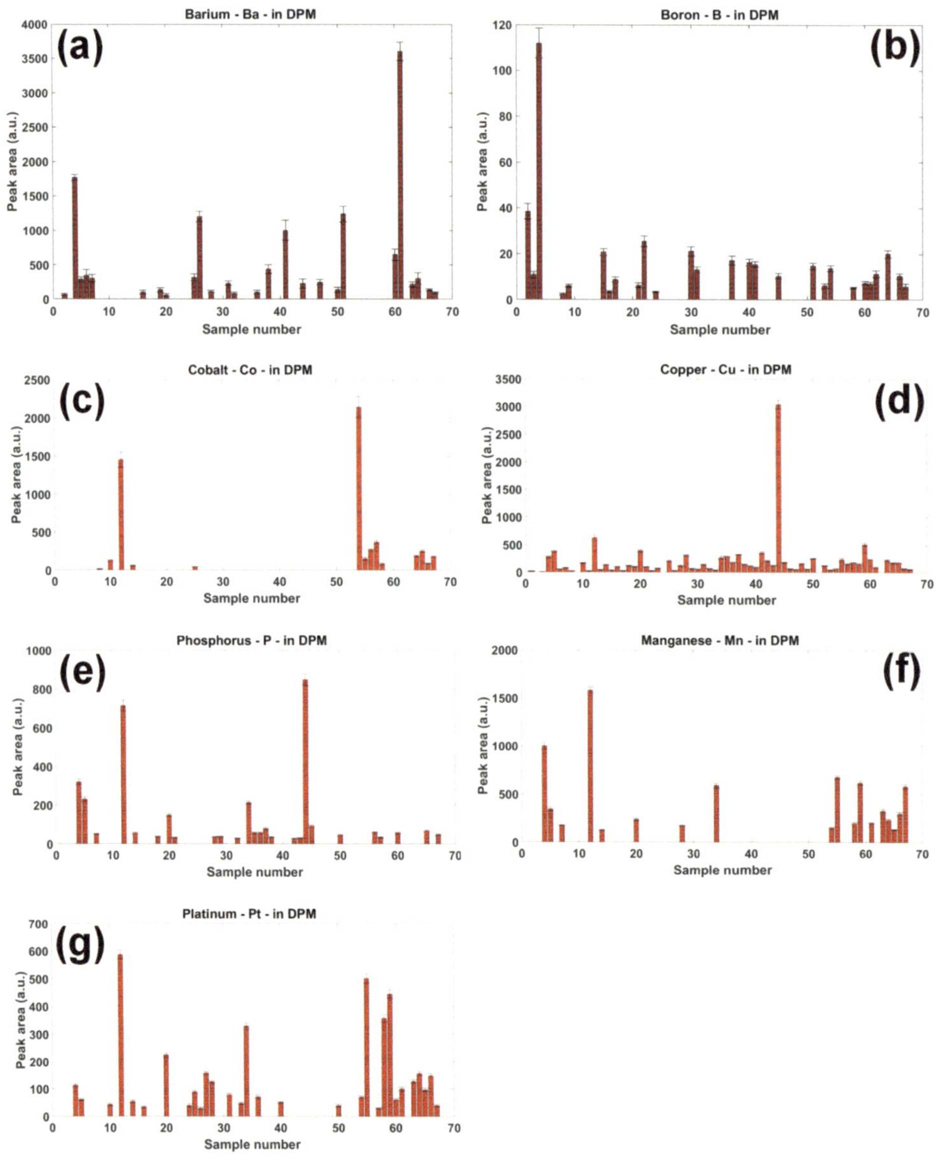

Figure 4.
Comparison of calculated integral values from LIBS optical emission spectra of barium (a), boron (b), cobalt (c), copper (d), phosphorus (e), manganese (f) and platinum (g) trace elements in diesel particulate matter, collected from different in-use passenger diesel engine vehicles.

comparison of LIBS signal of major and minor elements in DPM, we could classify the Barium as a trace element in DPM.

Boron spectral line: in **Figure 3(b)**, measured atomic emission from boron, doublet spectra line B I @ 208.88 nm and B I @ 208.95 nm, is shown. Here we selected six different samples with line emission from boron, where the LIBS signal is clearly visible. The comparison of calculated integral peak values is shown in **Figure 4(b)**. Samples with high content of boron are 4, 2, 22, 30, 15 and 64. Boron is in DPM present as trace element in 27 different samples.

Cobalt spectral line: optical emission from Co II @ 228.61 nm is shown in **Figure 3(c)**. Here, the relatively higher content of cobalt was measured in samples 54, 12, 57, 56, 65, 64, 67 and 55. From numerical calculation of Co II spectral line, signal from cobalt emission was measured in 14 different DPM matrices, as shown in **Figure 4(c)**.

Copper spectral line: this is shown in **Figure 3(d)** as ionic Cu II @ 224.70 nm in ultraviolet spectral range. High content has been measured in samples 44, 12, 59, 20, 5, 41, 37 and 28. The comparison of integral spectral peak calculated values is shown in **Figure 4(d)**. Copper is present in 62 different samples.

Phosphorus spectral line: phosphorus spectral signal from six DPM samples is shown in **Figure 3(e)**. The observed phosphorus ionic line P II @ 221.03 nm is present in UV spectral range. The compared results from numerical calculation of integral peak values are shown in **Figure 4(e)**. Phosphorus trace element has been measured in 26 different DPM samples. Strong LIBS signal from phosphorus element is in samples 44, 12, 4, 5, 34 and 20.

Manganese spectral line: atomic emission from manganese triplet Mn I @ 403.07 nm, Mn I @ 403.30 nm and Mn I @ 403.44 nm is shown in **Figure 3(f)**. From this figure, one can observe higher content of manganese in samples 12, 4, 55, 59, 34 and 67. Manganese trace element has been measured in 18 different DPM samples, and the comparison of integral values is shown in **Figure 4(f)**.

Platinum spectral line: atomic emission from platinum chemical element is shown in **Figure 3(g)**. Here, the spectral line Pt I @ 203.24 nm from eight different DPM samples is clearly visible. Platinum as the trace element was measured in 30 different samples. While most of the intense signal was recorded from samples 12, 55, 59, 58, 34, 20, 27 and 64. Compared integral values are calculated and shown in **Figure 4(g)**.

Analyte	Spectral line	Wavelength (nm)	Detected in/ total number of samples	Most pronounced signal in samples	Chemical element
Ba	Ba II	455.40	26/67	61, 4, 51, 26, 41, 60	Barium
B	B I	208.95	27/67	4, 2, 22, 30, 15, 64	Boron
Co	Co II	228.61	14/67	54, 12, 57, 56, 65, 64, 67, 55	Cobalt
Cu	Cu II	224.70	62/67	44, 12, 59, 20, 5, 41, 37, 28	Copper
P	P II	221.03	26/67	44, 12, 4, 5,34, 20	Phosphorus
Mn	Mn I	403.07	18/67	12, 4, 55, 59, 34, 67	Manganese
Pt	Pt I	203.24	30/67	12, 55, 59, 58, 34, 20, 27, 64	Platinum

Table 1.
Summary of detected trace elements and spectral lines. Number of samples with detected trace signal and samples with most pronounced signal.

In **Table 1,** chemical elements (analyte), spectral lines, number of samples with detected trace element and most pronounced signal from trace element in particular sample number are summarised and investigated.

4. Conclusions

In this research, we have investigated the trace chemical elements contained in diesel particulate matter. The particulate matter has been collected from in-use diesel engine passenger vehicles randomly from different vehicles models. Particulate matter has been analysed spectrochemically by means of a high resolution laser-induced breakdown spectroscopy (LIBS). The qualitative LIBS measurements reveal the presence of trace chemical elements such as barium, boron, cobalt, copper, phosphorus, manganese and platinum in diesel particulate matter. These trace elements were observed as optical emission of atomic or ionic spectral line emission in laser produced plasma. The spectral signal from each trace element was further numerically calculated as integral value of peak width line to obtain qualitative results. From LIBS analytical measurements and calculated signal profile, we can summarise that barium has been detected in 26 from 67 samples. Traces of boron have been detected in 27 samples, cobalt in 14 samples, copper in 62 samples, phosphorus in 26 samples, manganese in 18 different samples and platinum in 30 from 67 DPM samples.

From our previous research, we found out that minor chemical elements in diesel particulate matter are Si, Ni, Ti, K, Sr and Mo [28]. While major chemical elements C, Fe, Mg, Al, Cr, Zn, Na, Ca, O and H are forming the most important part of diesel particulate matter composition [26, 27].

All these major, minor and trace chemical elements contained in particulate matter are contributing to overall exhaust emission composition from in-use diesel engine passenger vehicles.

Finally, we can conclude that the laser-induced breakdown spectroscopy technique is very powerful method for qualitative and quantitative characterisation of DPM. It can almost instantly measure the major, minor and trace components of DPM and thus provide high resolution spectrochemical information about the chemical composition of diverse particulate matter matrices.

The presence of detected major, minor and trace chemical elements in DPM exhaust emissions from in-use diesel engine passenger vehicles can be related to the different processes. Those are: engine combustion process itself, engine health state, dysfunctions of some exhaust filtering device components, etc. These are linked with injected fuel type, fuel additives, engine lubricants, engine wear out, state of selective catalytic reduction devices, insufficient soot or PM filtering devices, engine conditions and quality of intake air.

Nevertheless, in the future, the quantitative characterisation of trace elements and calibration procedure would be an advantage for precise monitoring of different trace concentrations. That would help to better understand the trace content in diesel particulate matter.

Acknowledgements

The authors would like to thank the Austrian Science Fund—FWF (Fonds zur Förderung der wissenschaftlichen Forschung) for providing financial support. This study was funded with the grant number: FWF—P27967.

This work has been also supported by the COMET-K2 Center of the Linz Center of Mechatronics (LCM) funded by the Austrian federal government and the federal state of Upper Austria.

Authors would like to thank Dr. Maria Rusnak for the proofreading and for the valuable corrections.

Author details

Richard Viskup*, Christoph Wolf and Werner Baumgartner
Institute of Biomedical Mechatronics, Johannes Kepler University, Linz, Austria

*Address all correspondence to: richard.viskup@jku.at; viskup@gmail.com

References

[1] Noll R. Laser-Induced Breakdown Spectroscopy, Fundamentals and Applications. Berlin Heidelberg: Springer-Verlag; 2012. ISBN: 978-3-642-20667-2

[2] Miziolek AW, Palleschi V, Schechter I. Laser-Induced Breakdown Spectroscopy (LIBS), Fundamentals and Applications. Cambridge University Press; 2006. ISBN: 978-0-521-85274-6

[3] Cremers DA, Radziemski LJ. Handbook of Laser-Induced Breakdown Spectroscopy. John Wiley & Sons Inc; 2013. ISBN: 978-1-119-97112-2

[4] Hahn DW, Omenetto N. Laser-induced breakdown spectroscopy (LIBS), Part II: Review of instrumental and methodological approaches to material analysis and applications to different fields. Applied Spectroscopy. 2012;**66**(4):347-419. DOI: 10.1366/11-06574

[5] Noll R, Fricke-Begemann C, Brunk M, Connemann S, Meinhardt C, Scharun M, et al. Laser-induced breakdown spectroscopy expands into industrial applications. Spectrochimica Acta Part B: Atomic Spectroscopy. 2014;**93**:41-51. DOI: 10.1016/j.sab.2014.02.001

[6] Stehrer T, Praher B, Viskup R, Jasik J, Wolfmeir H, Arenholz E, et al. Laser-induced breakdown spectroscopy of iron oxide powder. Journal of Analytical Atomic Spectrometry. 2009;**24**:973-978. DOI: 10.1039/b817279j

[7] Schauer JJ, Kleeman MJ, Cass GR, Simoneit BRT. Measurement of emissions from air pollution sources. 2. C-1 through C-30 organic compounds from medium duty diesel trucks. Enviromental Science & Technology. 1999;**33**(10):1578-1587. DOI: 10.1021/es980081n

[8] Lough GC, Schauer JJ, Park JS, Shafer MM, Deminter JT, Weinstein JP. Emissions of metals associated with motor vehicle roadways. Environmental Science & Technology. 2005;**39**(3):826-836. DOI: 10.1021/es048715f

[9] Cheung KL, Ntziachristos L, Tzamkiozis T, Schauer JJ, Samaras Z, Moore KF, et al. Emissions of particulate trace elements, metals and organic species from gasoline, diesel, and biodiesel passenger vehicles and their relation to oxidative potential. Aerosol Science and Technology. 2010;**44**(7):500-513. DOI: 10.1080/02786821003758294

[10] Ntziachristos L, Ning Z, Geller MD, Sheesley RJ, Schauer JJ, Sioutas C. Fine, ultrafine and nanoparticle trace element compositions near a major freeway with a high heavy-duty diesel fraction. Atmospheric Environment. 2007;**41**(27):5684-5696. DOI: 10.1016/j.atmosenv.2007.02.043

[11] Kleeman MJ, Schauer JJ, Cass GR. Size and composition distribution of fine particulate matter emitted from motor vehicles. Enviromental Science & Technology. 2000;**34**(7):1132-1142. DOI: 10.1021/es981276y

[12] Packer AP, Sarkis JES, Giné MF, Santos ÉJ. High pressure Asher (HPA-S) decomposition of biodiesel samples for elemental analysis by inductively coupled plasma optical emission spectrometry (ICP OES). Journal of Brazilian Chemical Society. 2014;**25**:743-749. DOI: 10.5935/0103-5053.20140028

[13] Fontaras G, Karavalakis G, Kousoulidou M, Tzamkiozis T, Ntziachristos L, Bakeas E, et al. Effects of biodiesel on passenger car fuel consumption, regulated and non-regulated pollutant emissions over legislated and real-world driving cycles.

Fuel. 2009;**88**:1608-1617. DOI: 10.1016/j.fuel.2009.02.011

[14] Edlund M, Visser H, Heitland P. Analysis of biodiesel by argon–oxygen mixed-gas inductively coupled plasma optical emission spectrometry. Journal of Analytical Atomic Spectrometry. 2002;**7**:232-235. DOI: 10.1039/b111476j

[15] Gangwar JN, Guptab T, Agarwal AK. Composition and comparative toxicity of particulate matter emitted from a diesel and biodiesel fuelled CRDI engine. Atmospheric Environment. 2012;**46**:472-481. DOI: 10.1016/j.atmosenv.2011.09.007

[16] Agarwal AK, Gupta T, Kothari A. Particulate emissions from biodiesel vs diesel fuelled compression ignition engine. Renewable and Sustainable Energy Reviews. 2011;**15**:3278-3300. DOI: 10.1016/j.rser.2011.04.002

[17] Betha R, Balasubramanian R. Emissions of particulate-bound elements from stationary diesel engine: Characterization and risk assessment. Atmospheric Environment. 2011;**45**:5273-5281. DOI: 10.1016/j.atmosenv.2011.06.060

[18] Ferioli F, Puzinauskas PV, Buckley SG. Laser-induced breakdown spectroscopy for on-line engine equivalence ratio measurements. Applied Spectroscopy. 2003;**57**(9):1183-1189

[19] Ferioli F, Buckley SG. Measurements of hydrocarbons using laser-induced breakdown spectroscopy. Combustion and Flame. 2006;**144**(3):435-447

[20] Joshi S, Olsen DB, Dumitrescu C, Puzinauskas PV, Yalin AP. Laser-induced breakdown spectroscopy for In-cylinder equivalence ratio measurements in laser-ignited natural gas engines. Applied Spectroscopy. 2009;**63**(5):549-554

[21] Gross V, Kubach H, Spicher U, Schiessl R, Maas U. Laserzündung und Verbrennung im Ottomotor mit Direkteinspritzung. MTZ - Motortechnische Zeitschrift. 2010;**71**(7-8):532-539

[22] Buschbeck M, Buchler F, Halfmann T, Arndt S. Laser-induced breakdown spectroscopy for lambda quantification in a direct-injection engine. Spectrochimica Acta Part B: Atomic Spectroscopy. 2012;**74-75**:103-108

[23] Lopez-Quintas I, Mateo MP, Pinon V, Yanez A, Nicolas G. Mapping of mechanical specimens by laser induced breakdown spectroscopy method: Application to an engine valve. Spectrochimica Acta Part B: Atomic Spectroscopy. 2012;**74-75**:109-114

[24] Kiefer J, Li ZS, Alden M. Laser-induced breakdown spectroscopy in a partially premixed turbulent jet flame. Measurement Science and Technology. 2013;**24**(7):075205

[25] Hsu PS, Gragston M, Wu Y, Zhang ZL, Patnaik AK, Kiefer J, et al. Sensitivity, stability, and precision of quantitative Ns-LIBS-based fuel-air-ratio measurements for methane-air flames at 1-11 bar. Applied Optics. 2016;**55**(28):8042-8048

[26] Viskup R, Wolf C, Baumgartner W. Qualitative and quantitative characterisation of major elements in particulate matter from in-use diesel engine passenger vehicles by LIBS. Energies. 2020;**13**:368. DOI: 10.3390/en13020368

[27] Viskup R, Wolf C, Baumgartner W. Major chemical elements in soot and particulate matter exhaust emissions generated from in-use diesel engine passenger vehicles. Introduction to Diesel Emissions, Chapter 6. London, UK: IntechOpen; 2020. ISBN: 978-1-78984-035-3. DOI: 10.5772/intechopen.90452

[28] Viskup R, Wolf C, Baumgartner W. Identification of the minor chemical elements in the particulate matter exhaust emissions from in-use diesel engine passenger vehicles. Diesel and Gasoline Engines, Chapter 9. London, UK: IntechOpen; 2020. ISBN: 978-1-78985-248-6. DOI: 10.5772/intechopen.90760